T0296694

Cambridge Elements ≡

Elements in Organizational Response to Climate Change
edited by
Aseem Prakash
University of Washington
Jennifer Hadden
University of Maryland
David Konisky
Indiana University
Matthew Potoski
UC Santa Barbara

GOVERNING SEA LEVEL RISE IN A POLYCENTRIC SYSTEM

Easier Said than Done

Francesca Pia Vantaggiato
King's College London
Mark Lubell
University of California Davis

CAMBRIDGE
UNIVERSITY PRESS

Shaftesbury Road, Cambridge CB2 8EA, United Kingdom

One Liberty Plaza, 20th Floor, New York, NY 10006, USA

477 Williamstown Road, Port Melbourne, VIC 3207, Australia

314–321, 3rd Floor, Plot 3, Splendor Forum, Jasola District Centre, New Delhi – 110025, India

103 Penang Road, #05-06/07, Visioncrest Commercial, Singapore 238467

Cambridge University Press is part of Cambridge University Press & Assessment, a department of the University of Cambridge.

We share the University's mission to contribute to society through the pursuit of education, learning and research at the highest international levels of excellence.

www.cambridge.org

Information on this title: www.cambridge.org/9781009475945

DOI: 10.1017/9781009433594

© Francesca Pia Vantaggiato and Mark Lubell 2024

First published 2024

A catalogue record for this publication is available from the British Library.

ISBN 978-1-009-47594-5 Hardback
ISBN 978-1-009-43358-7 Paperback
ISSN 2753-9342 (online)
ISSN 2753-9334 (print)

Governing Sea Level Rise in a Polycentric System

Easier Said than Done

Elements in Organizational Response to Climate Change

DOI: 10.1017/9781009433594
First published online: March 2024

Francesca Pia Vantaggiato
King's College London

Mark Lubell
University of California Davis

Author for correspondence: Francesca Pia Vantaggiato,
francesca.vantaggiato@kcl.ac.uk

Abstract: How do polycentric governance systems respond to new collective action problems? This Element tackles this question by studying the governance of adaptation to sea level rise in the San Francisco Bay Area of California. Like climate mitigation, climate adaptation has public good characteristics and therefore poses collective action problems of coordination and cooperation. The Element brings together the literature on adaptation planning with the Ecology of Games framework, a theory of polycentricity combining rational choice institutionalism with social network theory, to investigate how policy actors address the collective action problems of climate adaptation: the key barriers to coordination they perceive, the collaborative relationships they form, and their assessment of the quality of the cooperation process in the policy forums they attend. Using both qualitative and quantitative data and analysis, the Element finds that polycentric governance systems can address coordination problems by fostering the emergence of leaders who reduce transaction and information costs. Polycentric systems, however, struggle to address issues of inequality and redistribution.

Keywords: climate adaptation, sea level rise, polycentric governance, San Francisco Bay Area, Ecology of Games

ISBNs: 9781009475945 (HB), 9781009433587 (PB), 9781009433594 (OC)
ISSNs: 2753-9342 (online), 2753-9334 (print)

Contents

1 Introduction: Climate Adaptation and Collective Action

Introduction

Climate adaptation will be a crucial global issue for the foreseeable future and includes adapting to climate vulnerabilities like sea level rise, wildfires, extreme heat, drought, and other impacts (Hinkel et al. 2018; IPCC 2014). But climate adaptation is "easier said than done" because it poses a new set of collective action problems that requires the evolution of polycentric governance arrangements. Even when policy actors are aware of the problem, solving the problem requires sustained cooperation and learning along with institutional changes that support on-the-ground adaptation strategies and projects. Climate adaptation is often treated as a private good, where the benefits and costs of adaptation accrue only to individual actors (Tompkins & Eakin 2012). However, treating climate adaptation as a private good is a false assumption – climate adaptation features important interdependencies where adaptation behaviors have social costs and benefits (Woodruff et al. 2020) and thus necessitates cooperation.

The core causes of collective action problems in climate adaptation can be found in vulnerability and adaptation interdependencies (Hummel et al. 2018). Vulnerability interdependencies occur when a climate impact in one jurisdiction has cascading effects on other actors and jurisdictions. For example, when coastal flooding occurs in one local jurisdiction, there are cascading infrastructure effects on transportation throughout the region (Madanat et al. 2019). Adaptation interdependencies occur when the adaptation actions of one actor increase or decrease climate risks for others. For example, when one local jurisdiction builds a sea wall or other type of coastal protection, the resulting hydrodynamic feedback may increase flood risks in other jurisdictions (Hummel & Wood et al. 2018). Such interdependencies exist in most climate adaptation contexts and require a regional or collective approach to governance. Due to climate change "lock-in" (Groen et al. 2022), the collective action problems of climate adaptation will occur even if carbon emissions are instantly reduced to zero.

In the context of the cooperation problems around climate mitigation, scholars have pointed out the importance of polycentric governance systems (Allan et al. 2021; Keohane & Victor 2011; Koski & Siulagi 2016). Ostrom argues that effective climate mitigation requires polycentric governance, which features "a complex combination of multiple levels and diverse types of organizations drawn from the public, private, and voluntary sectors that have overlapping realms of responsibility and functional capacities" (McGinnis & Ostrom 2012, p. 15). More generally, Ostrom et al. (1961; see also Carlisle and Gruby 2017) define polycentric governance as many formally autonomous units that consider each other's actions through a process of cooperation, conflict, and conflict resolution.

However, the notion of polycentric governance is rarely used as an explicit concept or theory in the context of climate change adaptation (Biesbroek & Lesnikowski 2018). This is puzzling because polycentricity permeates adaptation efforts all over the world, as they feature local action, experimentation, and policy networks fostering the emergence of trust and coordination among different stakeholders (Biesbroek & Lesnikowski 2018). Hence, our core research question is: *how do polycentric governance systems respond to the emerging collective action problems associated with climate adaptation?*

To answer this question, this Element advances both theory and empirical research in the context of climate adaptation. In terms of theory, we merge two frameworks: the climate adaptation planning cycle and the Ecology of Games Framework (EGF). The climate adaptation planning cycle describes how stakeholders involved in climate adaptation overcome barriers to move through the stages of understanding the problem, planning, and implementation, with feedbacks over time. As a general theory of polycentric governance (Berardo & Lubell 2020; Lubell 2013) the EGF analyzes how diverse policy actors participate across multiple policy forums where they deliberate and make decisions about interconnected issues. The capacity to solve collective action problems is a function of processes of learning, cooperation, and bargaining within this complex system. Thus, the barriers considered in the adaptation planning cycle are related to the capacity of a polycentric system to catalyze these key social processes.

For our empirical analysis, we use the case study of sea level rise in San Francisco Bay (SF Bay), California. Sea level rise is widely recognized as one of the most important climate impacts facing SF Bay, which is a highly urbanized region with significant population density, built environment, and critical infrastructure vulnerable to the combination of sea level rise and flooding from extreme precipitation events (Hummel et al. 2018). The political culture in SF Bay is generally progressive and concerned about climate change issues, and features a rich ecosystem of environmental policy institutions that is a hallmark of California's reputation for policy innovation. In response to concern about sea level rise, SF Bay has experienced a rapidly evolving polycentric governance system with new forums emerging across multiple levels of geographic scale (Lubell & Robbins 2021). This includes local level adaptation plans and projects by cities and special districts, as well as regional planning efforts led by government agencies seeking to coordinate adaptation efforts among all actors. Hence, SF Bay is in the midst of the adaptation planning cycle and facing significant governance barriers, which allows us to study the evolution of polycentric governance systems as it happens.

The overarching argument of the EGF is that effectively responding to new collective action problems requires the social processes of learning, cooperation, and bargaining that play out over time as polycentric systems move from awareness of a new collective action problem to institutional change and the implementation of on-the-ground climate adaptation policies (Lubell et al. 2021). Moving from one stage to the next requires overcoming various barriers to adaptation (Ekstrom & Moser 2014), which is why climate adaptation often becomes stuck in earlier stages. From the EGF perspective, moving through the climate adaptation planning cycle entails the evolution of polycentric institutions that overcome the transaction costs of searching for new policy agreements, bargaining over the various options, and monitoring and enforcing any agreements.

The next section will elaborate on our integration of the climate adaptation planning cycle with theories on polycentric governance, and lay out the main empirical research questions that will be the focus of different sections. We then provide more details about the case study and empirical research design, and summarize the overall plan of the Element.

Theory: Linking the Climate Adaptation Planning Cycle to Polycentric Governance

The adaptation planning cycle is a practical model that begins with understanding the problem, continues with identifying and planning for adaptation options, and ends with implementing and monitoring adaptation strategies. Moving through this cycle requires overcoming various barriers to adaptation, and adaptation planning often becomes stuck at earlier stages. However, climate adaptation planning does not explicitly consider the collective action problems involved with climate change adaptation, or the governance structure and processes in which the planning cycle is embedded. The EGF fills these theoretical lacunae by analyzing polycentric governance as a system where multiple actors participate in many different forums to deliberate and make collective decisions about interconnected collective action problems. To effectively overcome the barriers in the adaptation planning cycle, the polycentric system must facilitate the social processes of learning, cooperation, and bargaining.

The Climate Adaptation Planning Cycle

As an outgrowth of the earlier idea of adaptive management (Folke et al. 2005; Lee 2001), the adaptation planning cycle is a practical model designed to guide adaptation planning and decision-making. Ekstrom and Moser (2014) decompose the three main phases of Understanding, Planning, and Managing into different substages. For Understanding, the adaptation cycle includes the substages of understanding, gathering information, and defining the problem. In the context of climate adaptation, the understanding is often integrated in the context of climate impact analysis or vulnerability assessment, including scenarios linking climate impacts to higher or lower emissions pathways. More recent examples often analyze the inequitable distribution of the costs and benefits of climate change, and hence directly intersect with concepts of environmental and climate justice.

Once the problem is understood, the Planning phase involves developing, assessing, and selecting options. This is generally the phase of adaptation planning, which seeks to identify the various policy actions and on-the-ground projects needed to increase adaptive capacity and resilience with respect to priority climate impacts.

Finally, the Managing phase includes implementing the preferred adaptation options and monitoring and evaluating their impacts. This stage requires coordinated action by all the actors with the resources and authority to implement the actions, as well as developing metrics to monitor progress. In theory, the cycle adapts and begins again via monitoring feedbacks that identify the successes and failures of previous decisions and needed next steps.

The adaptation planning cycle is often depicted as temporally ordered, and a common analytical approach is to analyze how far along the adaptation planning cycle any case study has moved. However, progressing through the climate adaptation cycle is stymied by various adaptation barriers (Biesbroek et al. 2014) and governance challenges (Lubell 2017; Lubell

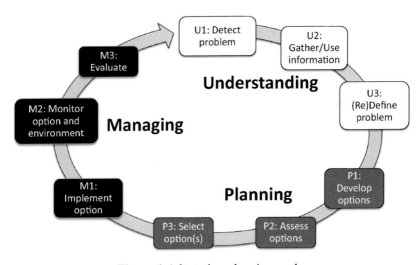

Figure 1 Adaptation planning cycle.

Figure from Ekstrom and Moser (2014) – reproduced with permission from Elsevier.

et al. 2021). One goal of the literature is to identify the most common governance barriers and think about potential solutions. The highest barriers to adaptation comprise various types of governance challenges, including institutional fragmentation and overlap (Ekstrom & Moser 2014). Another goal is to develop theoretical arguments about the individual level, social, and institutional processes that underpin barriers. For example, Moser and Ekstrom (2010) organize barriers into spatial/jurisdictional and temporal dimensions, where each barrier is analyzed relative to the position of a particular actor.

Stakeholder and community engagement are important aspects of both the research and application of the adaptation planning cycle (Pasquier et al. 2020). The adaptation planning cycle does not occur in a vacuum; the best practice is to include stakeholders and communities at all stages. Stakeholder engagement builds trust, increases legitimacy, and integrates diverse sources of knowledge. Community engagement strategies are also considered for increasing the procedural fairness of adaptation planning, and thus mitigating climate and environmental injustices (Dobbin & Lubell 2019; Dobbin et al. 2023). Especially in the context of community engagement, the adaptation planning cycle overlaps substantially with theories of collaborative governance (Ansell & Gash 2008; Emerson & Gerlak 2014).

Polycentric Governance as an Ecology of Games

However, in our opinion, the existing research on the adaptation planning cycle lacks an adequate analysis of how the cycle is embedded in polycentric systems. Even though many authors recognize its existence and importance (Folke et al. 2010), polycentric governance is often treated as a normative black box rather than subjected to a critical analysis of structure and function with respect to the effectiveness of adaptation planning. In other words, polycentric systems are often prescribed as the preferred institutional arrangement for adaptive environmental governance, relative to a more "centralized" or "monocentric" approach. This normative prescription ignores both the ubiquity and variance in polycentric systems, as well as the factors that might cause them to be more or less effective at solving collective action problems (Berardo & Lubell 2019).

To remedy these issues, we adopt the EGF as a conceptual framework. The EGF argues that polycentric systems are composed of policy actors participating in policy forums where they deliberate about a set of interlinked policy issues or collective action problems. The resulting constellation of interdependencies, which can be depicted and analyzed as a network, constitutes a system of policy games in which actors make strategic decisions. Policy actors are usually representatives of public or private organizations that are impacted by the decisions made in a specific system. Policy issues are social, economic, or environmental processes that are of interest to a certain number of policy actors and therefore shape the policy preferences of those actors. Policy issues often result from collective action problems or distributional conflicts. Policy forums are the decision-making processes in which involved actors deliberate and make collective choices about the policy issues (Lubell et al. 2022).

The "games" that give the framework its name are "policy games" (Long 1958), comprising the interactions between policy actors, forums, and issues, along with the institutional rules governing decision-making. The set of policy games can be considered a polycentric system and features structural attributes that can be measured in research design. Each of these components also has individual-level attributes, such as the social values of the actor, the type of issue, or the geographic scope of the forum.

The interactions among policy issues, actors, and forums play out over time in the context of a policy system. Policy systems are "geographically defined territories" that encompass multiple issues, multiple forums, and multiple actors interacting over time (Lubell 2013, p. 542). The interactions involve three key processes (also called "functions" in EGF parlance): (1) *learning* about causal drivers of policy issues and the features of the actors and policy forums; (2) *cooperation* in developing and implementing policy and using resources; and (3) *bargaining* over the distribution of the costs and benefits of policies.

The intersection between the adaptation planning cycle and EGF is most obvious in the "transaction cost" hypothesis, which draws from neo-institutional economics in arguing that policy forums will produce more cooperation if they reduce the transaction costs of searching for policy agreements, bargaining over the distribution of costs/benefits, and monitoring and enforcing the resulting policy agreements. Transaction costs can thus be conceptualized as the source of various barriers to adaptation. Hence, research in the EGF tradition has focused on various factors that may increase or reduce transaction costs, and thus create barriers or help overcome them.

Transaction costs may be influenced by the attributes and behaviors of individual actors, the structure of policy networks, and institutional arrangements within and between forums. For example, the *risk hypothesis* (Berardo & Scholz 2010) holds that policy networks provide access to different types of social capital depending on the level of risk actors face in a governance system. A recent contribution addressed some of the shortcomings of the "risk hypothesis" by developing a *multifunctional hypothesis*, which contends that polycentric systems must support different functions (learning, bargaining, and cooperation) at the same time and over time. Like the human brain, effective polycentric systems are likely to be functionally differentiated (Vantaggiato & Lubell 2023). There may also be tradeoffs between different functions, which requires balancing the costs and benefits of learning, cooperation, and bargaining across the system (Hamilton 2018).

Vantaggiato & Lubell (2023) find that forums whose participants embody higher levels of specialized knowledge of the policy issue make tangible progress even in the presence of high levels of conflict. Vantaggiato et al. (2023) show that policy actors with higher resources and authority can reduce transaction costs in social–ecological systems by providing ample connectivity across the social and ecological levels, whether or not the two are presently interdependent. In so doing, these actors future-proof the system against future governance or ecological challenges.

Empirical Research Questions

The empirical analysis will focus on four different empirical questions found at the intersection of the adaptation planning cycle and polycentric governance:

1. How do policy actors perceive sea level rise as a new collective action problem?
2. What are the perceived governance barriers and solutions to adaptation to sea level rise?
3. How do policy leaders catalyze policy networks for learning and cooperation to respond to sea level rise?
4. How do the emerging policy forums perform with respect to cooperation and learning?

Each of these questions plays an important theoretical role in how the social processes of learning, cooperation, and bargaining co-evolve with institutional change in a polycentric governance system. The first question focuses on how policy actors perceive the problem of sea level rise as a new collective action problem. As a problem like sea level rise becomes more salient, policy actors are more likely to feel that they are informed about it and spend a substantial amount of work effort addressing it. For climate change issues, psychological distance is an important concept – policy actors are more likely to engage with climate change issues that they construe as spatially, temporally, or socially proximate. These topics are tackled in Section 2.

The second question is inspired by the adaptation planning cycle's concern about identifying and overcoming barriers to adaptation. Even when agreement on the existence and nature of the problem begins to emerge, agreement on solutions remains difficult because actors are heterogenous in their organizational types, level of geographic scale, values, policy preferences, and capacity. Knowledge of the climate science and technical solutions won't suffice to overcome the barriers and move to the planning phase of the cycle (Vignola et al. 2017); adaptation needs leadership, political will, and broad public engagement. We address these topics in Section 3.

The third question focuses on the emergence of policy networks that respond to new collective action problems like sea level rise, where policy actors form new collaborative relationships to work together to learn about the problem and develop policy solutions. In these networks, centralization responds to the need for coordination of the heterogeneous preferences of involved stakeholders, making it possible for leaders to focus their efforts on stakeholders with high propensity for collective action, in order to activate them and enhance the probability of triggering action and change. Here, the required style of leadership facilitates connections between experts, communities, and other stakeholders (Meijerink & Stiller 2013; Vignola et al. 2017). In Section 4, we describe the function of leadership in the network of adaptation to sea level rise (SLR) in SF Bay.

The fourth question analyzes the emergence of new policy games. As the polycentric system changes, new "policy games" (i.e., policy forums) are created to address the problem, and the "big games in town" are central in the network and attract many actors. The extent to which those "big games" catalyze learning and cooperation has cascading effects throughout the polycentric systems. In Section 5, we compare the performance of seven "big games" in SF Bay, with different types of mission: from networking and information diffusion, to vulnerability assessment, to planning. We test whether actors with a preexisting agenda (environmental groups, nongovernmental organizations, and local governments such as special districts) systematically differ from more neutral actors (governmental agencies, academics, and experts) in their assessment of forum performance, and whether their assessments depend on their level of involvement in the polycentric system.

The study thus has important implications for our understanding of both the literature and the practice of climate adaptation, and opens new theoretical and empirical questions for future research. For example: what kinds of governance problems can polycentric governance systems solve? How should the adaptation governance literature incorporate political/distributional conflicts in its planning cycle theory? How to study adaptation planning processes in different governance systems? We provide some provisional answers to these questions in the concluding section.

Case Selection: Sea Level Rise in San Francisco Bay

Why sea level rise? Why San Francisco Bay? Given the level of coastal development around the world, sea level rise will be one of the most globally costly climate change issues (Hinkel et al. 2018). From a theoretical standpoint, sea level rise and climate adaptation are interesting because they pose new collective action problems to which existing polycentric governance arrangements must react. From an empirical standpoint, San Francisco is important given the global economic value of its coastal development and infrastructure. More importantly, San Francisco is a good case study because it embodies the challenges that many similar places around the world will have to contend with; while most stakeholders agree that sea level rise is an important issue, they are in the throes of creating new institutional arrangements, and thus studying this case provides an opportunity to analyze institutional change in motion.

Most existing literature on the governance of climate adaptation focuses on planning at local level and the challenges therein (lacking resources, difficulty identifying priorities, etc.) (Bednar et al. 2019; Ekstrom & Moser 2014; Moser & Ekstrom 2010), or how to bring stakeholders around the table (Huitema et al. 2016). Our study focuses on a region of the world where collaboration and coordination across levels are commonplace. In the SF Bay Area, the challenge is not bringing people around the table; as this Element will show, the challenge is getting them to agree on anything once they are at the table. San Francisco Bay is a representative case of polycentric governance in an area with "an abundance of government" and governmental agencies dedicated to addressing environmental issues (Vogel 2018).

Focusing on a "typical" case of polycentric governance system allows us to shed light on those challenges that aren't addressed by collaboration and coordination between interested parties, even in policy contexts where actors are "trained" in addressing policy issues in that way. In other words, our contention is that "if polycentric governance fails in the Bay Area, it fails everywhere," and that's because some of the challenges of climate adaptation are straightforwardly political challenges, which are likely to be valid beyond the specific context of the San Francisco Bay Area and to apply to most coastal cities around the world. For these reasons, this study can be considered a report on a "pilot project" on the governance of climate adaptation that is relevant to the whole world.

Research Design

The empirical data comes from a mixed method approach that includes qualitative data, quantitative surveys, and participant observation. The qualitative data consists of thirty-nine semistructured interviews with key informants carried out in 2016–17, which identified some of the main challenges associated with SLR adaptation in SF Bay along with sets of possible solutions.

Based on what we learned from the qualitative phase of the research, in 2018 we fielded an online survey to both governmental and nongovernmental governance actors (N=722, response rate 23 percent) involved in SLR adaptation in SF Bay (Lubell et al. 2019). The survey focused on respondents' concerns and priorities concerning adaptation to SLR, the collaborative relationships they maintain, and their assessment of the performance of the policy forums they attend. These three sets of data form the core of the analyses in the three empirical sections (Sections 3–5). We expand on the data collection process in the next section.

Our research design also incorporates a high level of participant observation that is still ongoing at the time of this writing. We have presented our results multiple times to stakeholder groups throughout the Bay Area, including formal meetings of the Bay Conservation and Development Commission (BCDC) and the Bay Area Regional Collaborative. One of the authors served on the leadership group for the ensuing BayAdapt regional sea level rise plan developed by BCDC, including working with a subgroup focused on environmental justice. This same author also serves on the Science Advisory Board for the Delta Science Program, which taps into a network that overlaps with the SF Bay and considers sea level rise and other climate adaptation issues. This ongoing policy engagement provides deeper insight into the qualitative and quantitative results, and allows us to continue observing the polycentric system as it changes over time.

Overview of the Sections and the Findings

In Section 2, we introduce the informants we interviewed at the outset of the study in 2016–17. We accompany these with data drawn from the online governance survey we conducted in 2018, and show that the concerns and perceptions of our thirty-nine interviewees are mostly mirrored in the concerns and perceptions of the 722 respondents to our survey. The section outlines the characteristics of survey respondents and provides an overview of their answers to our questions concerning their main concerns related to SLR, the expected impact, and how informed they feel about it, as indicators of the salience of SLR to SF Bay stakeholders.

In Section 3, we address the climate adaptation planning literature and explain why understanding its intersection with the polycentric governance literature is helpful to studies of adaptation governance: in a nutshell, this is because adaptation planning often occurs within the context of polycentricity and fragmented governance. Understanding the functioning and capabilities of polycentric governance systems in addressing collective action problems helps to identify the highest (and lowest) barriers to adaptation and intensify efforts accordingly. By investigating our interviewees' perceived barriers (and prospected solutions) to SLR in SF Bay, we find, in line with existing literature, that institutional barriers are by far the most mentioned, along with funding barriers and the scientific complexity of vulnerability interdependencies across the Bay. As for the solutions, institutional and political leadership were the most frequently mentioned ones. While our findings mostly mirror Ekstrom and Moser's (2014) qualitative study of five cases of adaptation across SF Bay, we identify an ongoing shift from the Understanding phase to the Planning phase of the planning cycle, suggesting that the governance system of SLR in SF Bay has moved forward in the several years between their investigation and ours.

In Section 4, we use data from our 2018 online survey to understand the structure of the policy network of stakeholders involved in addressing SLR in SF Bay, focusing particularly on the function of leadership and its explanatory power of network structure. Stemming from the multifunctional hypothesis introduced in Vantaggiato & Lubell (2023), which argues that climate policy networks embed multiple social processes and therefore different types of social capital at the same time, the section shows that leadership is instrumental in helping the polycentric system move from the Understanding to the Planning phase of the adaptation cycle, by "knitting together" local efforts in addressing SLR. Importantly, leadership is heterogeneous in terms of governance levels (featuring organization from the local to the federal level) but not in terms of types of actors – most network leaders are either governmental actors or knowledge experts (i.e., universities, consultants, and research-focused nongovernmental organizations). These actors represent 45 percent of our respondents.

In Section 5, we use our survey data to assess perceptions of policy forum performance in different types of survey respondents. We focus on seven "big games," that is, seven forums that were most attended and well known at the time of our research; 76 percent of our respondents attended at least one of these seven games. We then divide survey respondents into "neutral" and "partial." Neutral respondents include governmental actors and experts, whom we posit to be mainly interested in achieving coordination rather than furthering a specific predetermined agenda. Partial actors include advocacy groups, individuals, and special districts (independent, special-purpose governmental units that exist separately from local governments, with substantial administrative and fiscal independence, see Bollens (1986)), which we posit to be mainly interested in furthering specific agendas. We test their perceptions of the performance of the seven forums using multilevel regression models with crossed effects. We find that "partial" actors report lower satisfaction with forum performance than "neutral" actors, particularly in terms of impact and effectiveness of the forums. This suggests that the governance system does not fully meet the goals of "partial" participants. Combined with the findings in Section 4, which do not identify any "partial" actor in the leadership group, these findings suggest that the governance process – at the time of our observation – was dominated by actors whose main remit is producing and diffusing information. The relative disengagement of partial actors may leave the polycentric system unable to move along the phases of the planning cycle given the lack of buy-in from affected stakeholders.

In Section 6, we discuss our findings and their meaning for four bodies of literature: on the adaptation planning cycle, polycentric governance, collaborative governance, and for the theories of the policy process. We then conclude with some hypotheses concerning the evolution of the polycentric system for SLR and recommendations for future research.

2 Sea Level Rise Adaptation as a New Collective Action Problem

Introduction

Sea level rise is a globally important climate adaptation issue given the proportion of the population that lives in coastal regions and the associated intensive development of infrastructure and built environment (Hinkel et al. 2018). Like many other "world cities," San Francisco Bay, California, is vulnerable to sea level rise and associated flooding and has already experienced flooding during king tide and extreme storm events (Stacey et al. 2017). There is a high level of coastal development in SF Bay, including globally important infrastructure such as international airports, shipping ports, and information technology companies. San Francisco Bay also features wide variance in types and socioeconomic status of communities, which highlights the environmental justice and equity issues involved with sea level rise adaptation.

To set the stage for analyzing how SLR adaptation is proceeding in the polycentric governance system of SF Bay, this section provides descriptive data on four key concepts: policy engagement, the perceived impacts of SLR, the timing of SLR, and the cognitive frames our interviewees refer to when describing SLR as a problem. The next section will delve into their perceived barriers to adaptation to SLR, and perceived solutions.

As for policy engagement, SF Bay features a diverse set of organizations involved in SLR, some operating at the regional level, while many others are focused on the local level. These policy actors have a high level of variance in terms of their level of policy engagement with SLR, such as how informed they are and the level of effort they devote to adaptation planning. There is a core of actors at multiple levels of geographic scale, who are taking leadership on SLR adaptation, while many other actors are only cursorily aware and involved. The extent to which an actor is involved is mostly related to how SLR impacts their organizational goals and the geographic scope of their jurisdiction or attention.

The perceived impacts of SLR are related to the potential severity of the problem and when it is expected to occur, as well as how it may affect economic, social, and environmental resources. Policy theory typically refers to the salience of a particular policy issue as a driver of engagement (Baumgartner & Jones 1993). The research on environmental attitudes and climate change emphasizes the concept of psychological distance to consider how problems like sea level rise are construed in temporal, geographic, and social dimensions (McDonald et al. 2015). Whereas most of our respondents are very concerned about SLR, most fret about long-term consequences rather than short-term ones. Overall, however, our sample of respondents clearly sees SLR as a very salient problem for the Bay.

The data presented in this section combines the results of both the qualitative key informant interviews and the quantitative stakeholder survey. We introduced our research design in Section 1, and expand on data collection, interview coding, and survey administration in the next section.

Data Collection

We started our research process in 2016–17 by collecting qualitative data via thirty-nine semistructured interviews with a total forty-two key informants (listed in the next section). Interviewees were selected based on their involvement in the polycentric governance system of SF Bay. We gauged their involvement in the system in two ways: desk-based research of policy reports and organizations' websites, and participant observation of policy meetings concerning sea level rise in the Bay Area. Our involvement in the policy system also helped identify relevant actors or those who could point us toward other relevant actors. This allowed us to achieve very good levels of "exposure" to many of our initial informants (Small & Cook 2021) over a long period of time (from 2016 to the present day). Exposure (i.e., the time spent interviewing people, see Small (2009)) affords the researcher a deeper understanding of the motivations and evolution of their informants, in ways that enrich the analysis by, for instance, providing guidance concerning the expected direction of the correlation between the variables of quantitative analyses, and helping to develop expectations concerning the shape and the structure of the network of relationships that exist within the system.

We coded the interviews (more about the coding process in Section 3) according to three headings: problem perceptions (or frames), barriers, and solutions. In doing so, we were inspired by the literature on adaptation barriers and planning and its calls to match barriers with solutions (Eisenack et al. 2014), while accounting for the socioeconomic and political context where adaptation takes place. In this section, we outline the key cognitive frames emerging from our interviews, which tell us about the political context of SF Bay for climate change, while in the next section we expand on barriers and solutions and how they are connected in the minds of our interviewees.

To collect a sample of contacts for the quantitative survey, we used interviews, policy reports, and meeting minutes of regional and local governance processes that were ongoing at the time of the research. We obtained a large list of contacts comprising 3,087 individuals, collectively representing 623 organizations. Our sampling frame was purposively inclusive, because it was impossible to tell from the large lists which individuals and organizations were actively involved in sea level rise adaptation, versus just being casual observers who stopped being engaged after receiving some basic information. Hence, we did not expect a high response rate to this survey, and we included questions to measure the respondents' overall level of engagement with sea level rise adaptation. We also asked our survey respondents to invite individuals that they know deal with sea level rise in the Bay Area to obtain a personalized survey link from us. We received eighteen requests for survey links. We invited all individuals on our contact list to complete our online governance survey on June 25, 2018. We closed the survey on September 10, 2018. A total 878 respondents filled at least 49 percent of the survey (response rate 28 percent) while a total 722 respondents completed or partially completed all sections of the survey, for a response rate of 23 percent. In the remainder of this Element, we use data from the latter group (N=722).

A Note on Network Boundary, Unit Nonresponse, and Item Nonresponse Within the Survey

Boundary specification is one of the most difficult tasks in the analysis of so-called "system-oriented networks" (Nowell & Milward 2022), that is, networks that emerge to deal with complex policy issues but do not have an official designation as such. The study of emergent systems involves mapping involved actors and the relationships between them to inductively reconstruct the social networks binding them all. This means that we do not and cannot know whether our list of contacts was comprehensive of all the organizations and individuals involved in the polycentric governance system of adaptation to SLR in SF Bay. It most likely was not.

This means that our network analyses suffer from unit nonresponse: some actors are involved in the polycentric system, but we did not observe their ties because they did not answer our survey. We know this happened: our response rate is relatively low, and many survey respondents named collaborators who did not answer our survey. For the purposes of the network analysis in Section 4, we consider missing actors as missing at random (MAR); this means that their probability of being missing may be dependent on other observed variables (e.g., type of organization) but not on other missing data (Krause et al. 2020; Schafer & Graham 2002). For example, we know that local community-based organizations were less likely to fill out our network questionnaire, because at the time of our observation they were less involved in the policy space of SLR in SF Bay. This tempers our concern with the reliability of our network analyses in describing the main features of the polycentric system.

Further, our data suffers from item nonresponse. In network data, this corresponds to tie nonresponse, meaning that survey respondents may not have mentioned all of their collaborators for reasons of faulty recall. Additionally, we cannot measure the ties of actors who were named as collaborators by survey respondents but did not complete our survey. This is a less problematic type of missingness than unit nonresponse because it retains more information per actor. We decided not to impute missing ties (Krause et al. 2020) but to include a matrix of structural zeros for actors who were partially observed (i.e., somebody named them as collaborators but they did not fill our survey – this is a total of 284 organizations). This is for three reasons. First, as mentioned earlier, we do not know the "true" size of the network; imputation presupposes that the true size of the network is known. Second, most of the actors who were mentioned by survey respondents but did not themselves complete the survey are only mentioned once, and none more than twice, suggesting that they may not be pivotal actors in the network. Third, larger and more centralized networks are usually more robust against missing data (Smith & Moody 2013).

We recognize that unit nonresponse and item nonresponse to the network question in our survey limits the generalizability of our findings with respect to the full SLR adaptation network in SF Bay. Unfortunately, these response rate issues are not unique to our study but rather a ubiquitous and arguably worsening problem for all policy studies that rely on stakeholder or decision-maker surveys (Manfreda et al. 2008). In particular, we expect that survey respondents tend to have a higher level of involvement with sea level rise adaptation than nonrespondents and thus our observed network is biased toward the highly involved core and misses less involved, peripheral organizations. However, we believe the observed network still has enough variance in degree distribution, network structures, and actor types to anchor our empirical methods.

In what follows, we report on SF Bay stakeholders' engagement, impacts, and timing of SLR using both the themes that emerged in the interviews and the corresponding questions we fielded in the quantitative survey to check the findings of the interviews across a larger sample of stakeholders involved in the governance system. We close with an outline of the cognitive frames our interviewees refer to when describing their perceptions of SLR as a problem.

Interviewees and Respondent Characteristics

Our interviewees comprise primarily governmental and nongovernmental organizations, including some representing businesses and private actors (e.g., the Bay Area Council). Agencies from all levels of government, from federal to state (agencies with regional mandates for SF Bay, for example, the Bay Conservation and Development Commission), to regional (i.e., Bay-wide) to local agencies all feature in our interviewees' list (see Table 1) alongside environmental NGOs with Bay-wide scope.

Table 1 Count of interviewees by organization

	Organization	Number of interviewees from organization	Type of organization
1	Bay Conservation and Development Commission (BCDC)	6	State Agency
4	Marin County	2	County government
2	Bay Area Council	1	Private sector NGO
3	Greenbelt	1	Environmental NGO
5	San Francisquito Joint Powers Authority	1	Joint Powers Authority (local government)
6	San Francisco Airport (SFO)	1	Local government
7	Association of Bay Area Governments (ABAG)	1	Regional Agency
8	Assembly member office	1	Politician
9	Bay Area Regional Collaborative (BARC)	1	Regional Agency
10	Bay Institute	1	Environmental NGO
11	Baykeeper	1	Environmental NGO
12	Caltrans	1	State Agency
13	Center For Ecosystem Management and Restoration (CEMAR)	1	Environmental NGO
14	California State Coastal Conservancy	1	State Agency

Table 1 (cont.)

Organization	Number of interviewees from organization	Type of organization
15	East Bay Dischargers Association	1	Joint Powers Authority
16	Marin Transit Authority	1	County Agency
17	Metropolitan Transportation Commission (MTC)	1	Regional Agency
18	Natural Resources Agency	1	State Agency
19	Natural Marine Sanctuary	1	Federal Agency
20	Point Blue	1	Environmental NGO
21	Port of Oakland	1	Local entity
22	San Mateo County	1	County government
23	Save The Bay	1	Environmental NGO
24	Sonoma County Transportation Authority- Regional Climate Protection Authority (SCTA-RCPA)	1	County Agency
25	City and County of San Francisco	1	Local government
26	San Francisco Regional Water Board	1	Regional Agency
27	San Francisco Estuary Institute (SFEI)	1	Environmental NGO
28	San Francisco Estuary Partnership (SFEP)	1	State Agency
29	San Francisco Public Utilities Commission (SFPUC)	1	Local government
30	South Bay Salt Pond	1	Restoration project
31	San Francisco Bay Area Planning and Urban Research Association (SPUR)	1	Urban Planning NGO
32	US Army Corps of Engineers (USACE)	1	Federal Agency

Figure 2 reports the percentage of different types of organizations that answered our quantitative survey, along with counts of respondents by type. For consistency, we created a subset of the respondents who answered all the questions reported in this Section (Figures 1 and 3–7). These were 619 out of our total 722 respondents. Of these, fifty-four marked themselves as involved on their own behalf ("Own involvement" in Figure 1), not on behalf of an organization. As shown in Figure 2, most of our survey respondents work in education or consulting, in local government, or for a nongovernmental organization (NGO). In previous work (Vantaggiato & Lubell 2023) we have split consultants from respondents who work in education or other research-based organization. In this Element, we merge those categories as these two types of actors perform the same function in the polycentric system: providing expertise and diffusing information. Educational institutions such as universities and research-based NGOs provide the scientific basis for SLR adaptation, especially in developing models for where flooding will occur under different scenarios of climate change, sea level rise, and storm events. Scientific organizations communicate this knowledge by directly participating in different forums.

Further, local governments are crucial actors who are often on the frontline of SLR and have land-use authority to implement on-the-ground adaptation strategies. The SF Bay Area is comprised of nine counties, and within those counties are 101 cities including the consolidated city-county of San Francisco. California is a "home-rule" state, which designates land-use authority to local governments. These local governments are usually

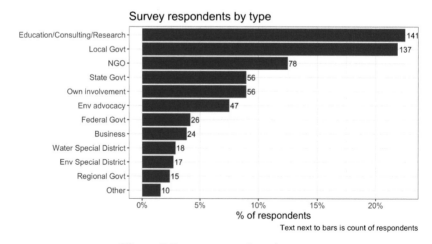

Figure 2 Survey respondents by type.

concerned with building coastal flood protection projects to mitigate flooding associated with SLR.

The SF Bay Area features many community-based groups and NGOs focused on environmental or social issues. Some of these community-based groups specialize on climate change, but most of them consider climate change and SLR as part of a broader portfolio of issues of concern. The community-based organizations are usually the main advocates for environmental justice concerns, and many of them represent disadvantaged and socially vulnerable communities. There are some emerging conflicts between the ecological goals of environmental groups, and the social goals of community-based organizations (Gmoser-Daskalakis et al 2023).

Special districts also play an important role in providing local infrastructure in the Bay Area, although the quantitative survey did a better job of measuring and representing them. For example, there are flood control districts, water management districts, wastewater districts, transit districts, and so on. The flooding associated with sea level rise may directly compromise the ability of these special districts to provide infrastructure services.

Federal, state, and regional government agencies are key actors in the polycentric system because of their political authority and policy resources. California has high institutional capacity, including state agencies with regional jurisdiction like the BCDC, which must issue permits for any land-use project within 100 feet from the shoreline of the Bay. BCDC also has an extensive planning division and has emerged as one of the leading actors for coordinating SLR adaptation in SF Bay. But other regional agencies also play important roles in terms of permitting authority and sources of funding for transportation and coastal infrastructure projects.

Despite having the policy resources of the US government, federal agencies play a less prominent role in SF Bay, mainly driven by a nexus with previous environmental policy responsibilities. For example, the US Fish and Wildlife Service (USFWS) manages an extensive complex of national wildlife refuges in the South Bay, which are both vulnerable to SLR and have a long history of wetland restoration that is now considered a crucial "green infrastructure" strategy for SLR adaptation. The USFWS also leads the SF Bay Joint Venture, which is a regional collaborative governance initiative focused on restoration of the Pacific flyway. As in all coastal regions, the US Army Corps of Engineers has important permitting authority for coastal development as well as a source of funding and expertise for major infrastructure projects, including both the classic "gray" infrastructure such as sea walls and levees and "green" infrastructure such as restored wetlands.

Figure 3 outlines the geographical distribution of the collaborative activities of our survey respondents based on the so-called Operational Landscape Units (OLUs). The OLU framework – developed by the San Francisco Estuary Institute (SFEI) – divides the Bay shoreline into thirty distinct geographic areas that share common physical characteristics and potential adaptation strategies (Beagle et al. 2019). OLUs cross traditional jurisdictional boundaries of cities and counties but adhere to the boundaries of natural processes like tides, waves, and sediment movement. Since their conception, OLUs have been adopted by key state and regional agencies, for example, e.g. BCDC, to frame the governance of SLR in the Bay. In our survey, we asked respondents to identify the OLUs they worked in by clicking on an interactive map of the OLUs that included key geographic features and place names. As can be seen in Figure 3, most of our respondents come from the central and North Bay, with the South and East Bay being overall less well-represented.

Policy Engagement with Sea Level Rise Adaptation

The online survey asked respondents specific questions about information and professional involvement. On the one hand, we wanted to gauge our respondents' perceptions of how well-placed they were to understand how to deal with SLR; on the other, we wanted an overview of their level of involvement in the polycentric system. Figure 4 shows that most policy actors consider themselves somewhat informed or well-informed about SLR, with no meaningful differences between their reported understanding of short versus long-term consequences (in fact, we had to jitter the points in the plot so that they would not overlap completely). The very few "not-informed" respondents reflect the correlation between survey response and knowledge; uninformed people are not interested in answering the survey. A significant portion of the sample considers themselves well-informed, and the overall pattern suggests a strong knowledge basis for SLR is present in the system.

However, while the extent of professional involvement is positively correlated with the level of information (Pearson's R=0.44), professional involvement appears to lag behind knowledge. Most survey respondents report SLR as being part of their work, with mostly occasional and sometimes routine involvement. Fewer respondents report SLR as a major aspect of their work, and most of them have been working on it for a few years (see Figure 5). The apparent lag between

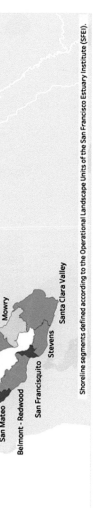

Figure 3 Distribution of shoreline segments that survey respondents work on in their SLR-related governance activities.
Figure from Lubell, Vantaggiato, and Bostic (2019) – reproduced with permission from author Darcy Bostic.

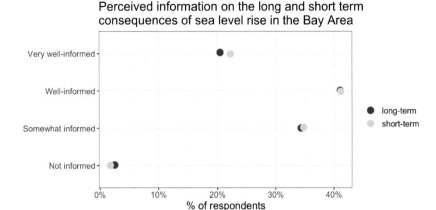

Figure 4 Survey data on information.

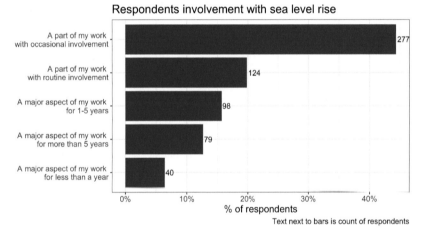

Figure 5 Survey respondents' involvement with the governance of SLR.

information and professional involvement echoes the ubiquitous gap between knowledge and behavior found in social psychology (Ajzen et al. 2011) and theories of environmental behavior (Bockarjova & Steg 2014; Davis et al. 2009). Many of these policy actors are tasked with working on the many different types of environmental collective action problems affecting SF Bay, most of which intersect with climate change and SLR adaptation. The potential lag between knowledge and engagement reflects the overall argument of this Element, that becoming aware of a new collective action problem is much easier than doing something about it.

The Psychological Distance of Sea Level Rise: Impacts and Timing

For a polycentric system to respond to a new collective action problem, it must be a salient issue for policy actors and the broader public (Baumgartner & Jones 1993). Theories of environmental behavior usually argue the perceived severity of the problem is a driver of behavior, which aligns with Ostrom's (1990) argument that perceived decline in common-pool resources is a motivator for the development of new institutions. In the climate change attitude literature, salience and perceived severity are related to the concept of "psychological distance," which argues that people are more likely to take action when climate change issues are more psychologically proximate in time, geographic space, and social categories (McDonald et al. 2015). The psychological distance of climate adaptation issues like SLR is related to when impacts are expected to occur, and the types of social and ecological systems that might be affected.

The quantitative stakeholder survey asked several specific questions about the psychological distance of SLR. Figure 6 shows that most policy actors are concerned or very concerned about SLR, but more about the long-term consequences rather than the short-term consequences. The policy narrative very often describes SLR as a "slow-moving" natural disaster, and many of the modeling scenarios focus on 2050 or 2100 as salient temporal milestones to frame decision-making.

However, despite higher concern for the long-term consequences, Figure 7 shows that the majority of SLR policy actors believe that SLR impacts have already started to occur. This is because the SF Bay Area regularly

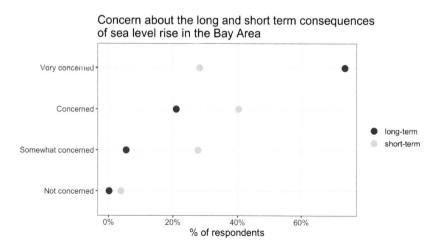

Figure 6 Survey respondents concern about SLR.

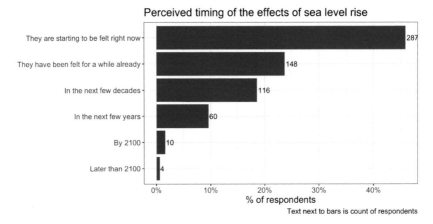

Figure 7 Survey respondents timing of impacts.

experiences coastal flooding during high ("King") tide events, and winter storm events especially during wet years. At times, such storm events lead to spectacular flood impacts such as the inundation of major roadways like Highway 37 or levee breaches in the Delta (Shilling et al. 2016). While such flood events cannot be wholly attributed to SLR, it does exacerbate the severity of these short-term events. SLR can be thought of as adding wood to the fire, with the short-term events providing the spark. Psychologically, it is easy to understand how policy actors may start associating the short-term flooding events with the slow change in the underlying risk parameters driven by SLR.

The impacts of SLR become more actionable when they are perceived to affect critical infrastructure in urban systems. Figure 8 shows that policy actors are most concerned about impacts to transportation and wastewater infrastructure. The SF Bay region is densely urbanized with many bridges, highways, airports, and public transportation routes located at low elevations near the coast. In a region that already experiences significant traffic congestion, coastal flooding can have significant cascading effects that reflect the vulnerability and adaptation interdependencies involved with SLR as a collective action problem (Madanat et al. 2019). Wastewater, stormwater, and water supply infrastructure are also often at low elevations (Hummel et al. 2018) and flooding may compromise the ability of these critical infrastructure systems to provide daily public goods.

Policy actors are also concerned about impacts to ecosystem and disadvantaged communities. These concerns reflect an overall tension between

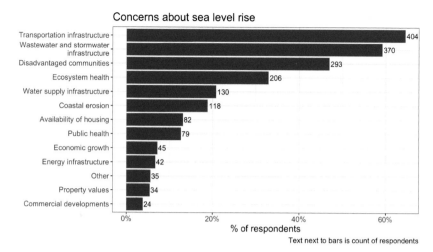

Figure 8 Survey respondents top concerns about the impacts of SLR.

environmental and social goals as emphasized by environmental groups and community-based organizations respectively. Environmental groups are concerned that flooding may overwhelm critical coastal ecosystems, which are already severely degraded in the SF Bay Area due to historical development. There are many threatened or endangered coastal species in the SF Bay, and coastal wetland and habitat restoration has been a key environmental priority for decades. In contrast, many community-based groups are more concerned with social and economic vulnerability in disadvantaged communities where coastal flooding has already affected local infrastructure (Meadows 2021). However, at this time the environmental and social priorities of SLR are not diametrically opposed in a way that shapes strong advocacy coalitions (Jenkins-Smith & Sabatier 1994). For example, most environmental groups would not deny the importance of environmental justice issues and most community-based groups would not deny the importance of coastal ecosystems. But there is a simmering tension between these environmental and social values that influences policy preferences (Gmoser-Daskalakis et al. 2023); while environmental groups prioritize conservation, ecosystem health, and habitat protection, community-based organizations prioritize flood protection for vulnerable groups. At the time of our observation, these tensions had not reached maturity and were not referred to any specific projects; yet given scarce resources to fund SLR projects, the question of which projects would be prioritized loomed large.

Understanding SLR as a Problem: Frames

The qualitative interviews revealed a more inductive set of problem frames that have clear connections to problems of learning, bargaining, and cooperation. The most frequently mentioned problem frame is that SLR requires regional coordination. SF Bay stakeholders are generally aware that increasing regional adaptive capacity requires coordinating local actions rather than local jurisdictions implementing independent strategies that do not consider vulnerability and adaptation interdependencies. While coordination and land-use are less frequently mentioned frames (see Figure 9), they are interwoven with the regional dimension of the problem. The fact that adaptation to SLR has land-use implications requires the close involvement of local authorities into any decision-making concerning what to build and where, increasing the relevance of institutional fragmentation as a challenge, since California is a very decentralized, multilevel system of governance (Henry et al. 2011).

On the more positive side, stakeholders also perceive SF Bay to have a favorable political climate, insofar as the political culture of the Bay Area is generally aware of climate change as a problem without a strong culture of denial. Furthermore, nearly half of the interviewees consider SLR not only a climate adaptation problem but also an environmental justice issue.

The key takeaway for us is that this is a "self-aware" polycentric governance system – stakeholders acknowledge climate change as well as institutional fragmentation and the necessity to coordinate to address the issue. This is important because it shows that, differently from other places around the

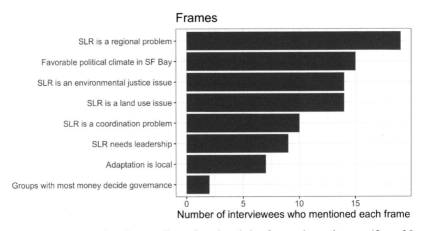

Figure 9 Frames of understanding of sea level rise for our interviewees (from 39 qualitative interviews).

world (Kammerer et al. 2021), in California not much convincing is needed to get stakeholders to "sit together at the table" to devise policy solutions.

Conclusion

This section outlined our data collection and the characteristics of the actors involved in the governance of SLR in the SF Bay Area via a combined analysis of our interviews (2016–17) and survey respondents (2018), as well as their overall engagement in the system, their perceptions of the main impacts of SLR, and their own level of information on SLR as an environmental issue. Overall, our conclusion is that SF Bay stakeholders are aware of SLR as an environmental issue and agree on the urgency to address it. Most of those we interviewed and surveyed are actively involved in the governance system, though only a third or so (i.e., those who said SLR is a "major aspect" of their work, see Figure 4) are heavily involved. Therefore, while our sampling approach has probably captured the most involved core of the polycentric system, there is variation in our respondents' level of involvement and, as shall be seen in the next sections, their perceptions of the ability of the polycentric system to address SLR. Overall, however, they feel well-informed about the impacts of SLR. The majority is very concerned about the future impacts of SLR, particularly on different types of infrastructure – this makes sense as SF Bay is a heavily urbanized metropolitan area home to over 8 million people.

Our interviewees and respondents belong to all levels of government as well as civil society, the private sector, and advocacy organizations. They see the regional dimension of the problem and understand that addressing it will require coordination and compromise. On this background, the governance context of the Bay Area appears as uniquely suitable to address the complex challenges of climate adaptation collaboratively and successfully. Yet even in such a favorable policy context, addressing SLR is far from a "done deal." The next section presents the findings of the interviews in more detail and shows the kinds of barriers perceived by our interviewees and the solutions they envisage. We find that institutional barriers (Young 2006) feature prominently among the core barriers that need addressing in order to successfully adapt to SLR in SF Bay.

3 Climate Adaptation Barriers in Polycentric Systems

Introduction

While research on the adaptation planning cycle typically identifies collective action and institutional fragmentation as barriers (Biesbroek et al. 2014; Moser & Ekstrom 2010), it rarely explicitly relates them to the polycentric structure of governance systems (Lubell et al. 2021). Furthermore, recent research in this field calls for a research agenda that goes "beyond describing and enumerating barriers towards explaining them" (Eisenack et al. 2014). This section heeds to this call by coupling investigation and description of the barriers to adaptation with the peculiarities of polycentric governance systems.

In this section, we investigate the barriers to adaptation to SLR in SF Bay as perceived by our interviewees. Importantly, in our interviews we also asked informants for their opinion on the solutions needed to overcome the barriers, thus responding to another call in Eisenack et al. (2014) to consider barriers to adaptation using actor-centered approaches. We categorized both barriers and solutions emerging from our qualitative interviews according to five categories, which dovetail those used in Ekstrom & Moser (2014): governance/institutional, scientific, political, financial, and attitudinal barriers.

The results of the qualitative analysis of the interviews show three things:

1. There is higher congruence concerning perceived barriers to addressing SLR, than concerning the potential solutions.
2. The core set of barriers that tend to co-occur across interviews comprise "institutional fragmentation," "vulnerability interdependence," and "funding." These speak to the three core pillars of the EGF: Institutional fragmentation speaks to the need for coordination and *cooperation* between stakeholders, vulnerability interdependence underscores the importance of *learning* about the risks to identify solutions, and (lack of funding) speaks to the difficulties of *bargaining* to decide what to build and where to protect SF Bay from SLR.
3. Although congruence in solutions is lower, interviewees expressed a strong demand for institutional and political leadership; thus, solutions related to demonstrating (governance and political) leadership "hit" many barriers at once.

Linking Adaptation Barriers and Solutions to Polycentric Systems and the EGF

The core idea of adaptation planning cycle frameworks is that climate adaptation must progress from problem identification, to planning, to implementation and adjustment in response to monitoring over time. Progress along the adaptation planning cycle is stymied by barriers of various sorts, which may be different for each stage (Biesbroek et al. 2014; Moser & Ekstrom 2010). Our argument is that such barriers are related to the learning, cooperation, and bargaining processes that are necessary for a polycentric system to respond to new collective action problems like SLR.

Polycentric systems have characteristics that affect how actors form their perceptions of the barriers and conceive potential solutions. Polycentric governance systems possess different but interdependent foci of discussion/deliberation, each focused on one aspect of the policy problem. The foci are interconnected by actors, who both conceptually, through the ways they understand the problem, and factually, by maintaining collaborative relationships with disparate actors and attending different policy forums, establish the links between the different facets of the policy problem. By taking part in the polycentric system, actors can observe the barriers as they manifest across different parts of the system, as well as the progress occurring to address them. They can use the knowledge gained in one process to steer outcomes in another, and/or envisage solutions that go beyond any specific case study or location to encompass the whole system.

We find our interviewees perceive a diverse range of barriers, many of which are related to the adaptation planning framework and were also found in previous work by Ekstrom and Moser (2014). Specifically, Ekstrom and Moser (2014) examined five case studies of adaptation to SLR and/or heat in the Bay Area. They carried out interviews to investigate the barriers to adaptation experienced by those involved. They found that governance/institutional barriers were the most prominent and concluded that this was due to the early stage of the adaptation cycle: The five cases were all still mostly in the understanding phase, where actors are focused on understanding their vulnerabilities and defining the problem in order to address it.

Ekstrom and Moser (2014) predicted that, as the cases moved through the cycle and into the planning phase, other types of challenges would become more poignant than they were at the time of their study, including lack of funding, institutional fragmentation, lack of a plan of action for adaptation planning, lack of vision, lack of leadership, and lack of governance structures to make decisions. As will be seen in this section, our findings echo their predictions quite closely, suggesting that we observed the governance system for SLR in SF Bay as it was moving from the understanding to the planning phase. Considering that the governance system for SLR in SF Bay finds its roots in policy forums organized starting in the late 1990s (Lubell & Robbins 2021), however, we may conclude that the progress of the governance system to the later stages of the planning cycle has not been rapid. This comports with the core argument of this Element, which is that adaptation barriers raise the transaction costs (search, bargaining, monitoring, and enforcement) of collective action. As a result, climate adaptation may become stuck at the earlier stages of the adaptation planning cycle and progress may not be rapid enough to keep pace with even "slow-moving" problems like SLR.

In this section, we show that the potential solutions to SLR envisaged by our interviewees are even more numerous and diverse than the barriers. In general, there is more agreement on the existence of SLR as a problem and less agreement concerning the solutions that should be implemented. It is in debating the solutions that the costs and benefits of adaptation become most salient to policy actors, and because policy solutions distribute costs and benefits in different ways, they must effectively bargain for a portfolio of solutions. It is here that the transaction costs of SLR adaptation are the highest, agreement is less likely to emerge, and the evolution of polycentric governance systems gets stuck in earlier stages of the adaptation planning cycle.

Coding of the Interviews

The interviews were semi-structured, with six questions asked of all interviewees while also allowing for follow-ups on the specific points raised by each interviewee. The six questions common to all interviews concerned, first, the interviewee's background and how they got involved in climate adaptation and SLR issues in SF Bay. Second, they were asked for their perceptions of the type of problem SLR is for the SF Bay – these answers formed the bulk of our frames categorization as reported in Figure 9 of Section 2. Third, they were asked about what they think are the main vulnerabilities of SF Bay to SLR. These answers form the bulk of the questions concerning impact and concerns as per Section 2. Fourth, interviewees were asked about what they think the main barriers to adaptation are in SF Bay. Fifth, they were asked about how to address them (i.e., their preferred solutions). The sixth and final common question to all interviewees asked them to recommend someone else to talk to, whose perspective may be valuable to the study. The themes emerged from the fourth and fifth questions concerning barriers and solutions form the core of the qualitative data discussed in this section.

We coded the qualitative interviews using the Discourse Network Analyzer (DNA) software (Leifeld 2020). Although originally conceived to track policy debates and political polarization (Leifeld 2013), the DNA software can be used to do an Nvivo-style type of analysis including taking notes sequentially. The software[1] allows for the creation of a database of documents (text files) and for multiple coders to work on the same database, using color codes to associate statements (and the interviewee metadata, e.g., name, organization, etc.) to specific concepts. The software allows the researcher to then export the interviews as a dataset linking individuals to their coded statements for further

---

[1] We used version 2.0 beta 21 of the DNA software. The software has now reached version 3.0.10, see https://github.com/leifeld/dna/releases (last accessed 11 August 2023).

manipulation in the R statistical software for the purposes of network analysis or, as in our case, visualization.

We used the capabilities of DNA to color-code interviewees' statements according to the five broad categories emerging from the literature on barriers to adaptation: institutional, political, scientific, financial, and attitudinal (Ekstrom & Moser 2014). We found that these five headings captured the core of the barriers expressed by our interviewees well. Since the five themes are very broad, however, we complemented their coding with additional descriptive labels of the concepts or concrete items that the interviewees associated with the five headings.

Figure 10 lists the barriers that emerged from the coding of the interviews. In Figure 10, the bar chart indicates how may interviewees mentioned that specific challenge. This approach has the benefit of clarifying how shared any given barrier is amongst our interviewees, and the downside of masking the interrelations between the various concepts as expressed by our interviewees. For example, jurisdictional fragmentation is the overall most mentioned challenge (see Figure 10). Yet, interviewees rarely mentioned it on its own. More often, interviewees linked each barrier to other barriers, for example. political leadership and inequities – which we coded accordingly.

To understand our coding approach to the interviews, consider the excerpt below, from an interview with a representative of an environmental NGO:

> More globally the challenge in the Bay Area, or regionally, is that we have 76 cities [on the shoreline] or something like that, and nine counties in the Bay and a lot of them have very differing perspectives on the level of control they want to have and the degrees to which they value sustainability, and that seems to give rise to inequities in that it's the wealthier cities that have the time and resources to commit to these issues and have the flexibility of rejecting potential development. Some of the less wealthy communities probably would like as much development as they can. A regional approach is definitely the highest priority, and then BCDC or a comparable agency having the political will to impose development restrictions and planning objectives for the region. (Interview 15)

In this excerpt, the interviewee links institutional fragmentation to anxieties about local control, the widespread income inequalities in SF Bay, the competing pressure of development, and the necessity of a coordinated regional approach, spearheaded by an agency or other actor displaying the necessary political leadership. Hence, we coded this single excerpt according to all of these themes: barriers include fragmentation, local control, inequality, and competing pressures; solutions include a regional approach and political leadership. In the following excerpt, a county official describes the tension between

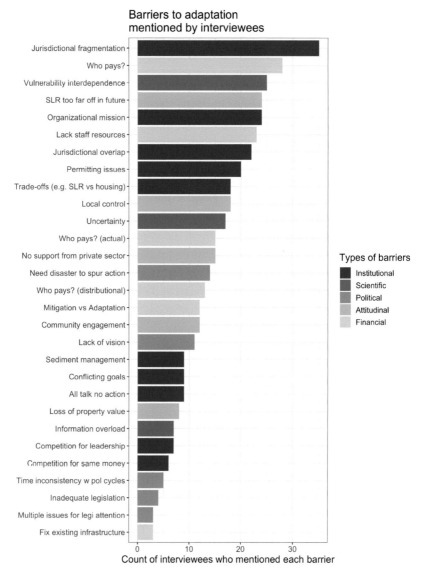

Figure 10 Barriers identified by our interviewees.

coordination and local control (a traditional preference for local control in California has been documented early on by Ostrom et al. (1961)). in starker terms:

> I mean, fundamentally, we have an abundance of government in the Bay Area, it's 100 plus cities and nine counties, and a lot of these issues are fundamentally land use issues, and land use is local jurisdiction, that's their

bread and butter and they're very, very hesitant to give up any control or decision making authority over any element of local land use. (Interview 8)

Others (primarily governmental actors) link institutional fragmentation to the difficulties of coordinating across departments of governmental agencies, pointing to organizational silos as the main obstacle to overcome. Yet other interviewees mention jurisdictional fragmentation as something that increases the sheer complexity of the task of dealing with SLR. In the words of a regional agency representative:

> The thing I struggle with in the Bay Area, is there are a lot of different tables which you're going to hear about, where stakeholders are coming together, I am involved in all of those tables basically, it's a complicated region, and it has a lot of government, a lot of high nonprofit capacity, we're really spread out. (Interview 9)

Another interviewee said SLR is "too big a problem. It's beyond most people's abilities to grapple with it, I think. There are so many stakeholders" (interview 24). Thus, while for all interviewees institutional fragmentation is a feature of the system that prevents coordinated action, some think of it descriptively as adding to the sheer complexity of the governance system; others link it to cross-level political tensions (local government's reluctance to delegate decision-making authority to higher levels of governance), while others mention intra-institutional silos. As another example, financial barriers (the second most mentioned type of barrier) included items ranging from restrictions on budget use at local level to lack of federal funding, to the trade-off between spending on mitigation and spending on adaptation.

Finalizing the descriptive labels required several rounds of re-reading and refining of the coding of all interviews to arrive at a list of barriers and solutions that was both comprehensive and manageable. While our visualization approach in Figure 9 does not fully render justice to these nuances, later in the section we link barriers to solution via a co-occurrence matrix showing which barriers and solutions tend to be mentioned together across interviews and may thus be more palatable complements to stakeholders in SF Bay.

Results: Barriers

The prominence of jurisdictional fragmentation as a barrier to adaptation resonates with most research on climate adaptation (Eisenack et al. 2014). Funding is the second most mentioned barrier. The funding barrier ("Who pays?" in Figure 9) results from the summation of two other headings in the bar chart: "Who pays? (actual)" and "Who pays? (distributional)." We use the former to indicate interviewees' comments that referred to the actual lack of sufficient funding (e.g., interview 3: "I think our main problem is going to be our limited dollars") and/or lack of appropriate funding pots and/or of federal support (e.g., interview 25: "as it becomes more and more common and there's like hundreds and hundreds of communities that are being hit there won't be the capacity of the federal government to take care of all of them").

We use the "Who pays? (distributional)" heading to indicate interviewees' concern about the distributional implications of paying for adaptation, given the high levels of inequality existing across local communities in the Bay Area: some of the most vulnerable municipalities have the lowest fiscal capacity. This is expressed in no uncertain terms by several interviewees, for example, interviewee 27:

> But then you go down to the San Bruno Creek zone, adjacent to the SFO airport, which has minimal flood control capacity, and a lot of lower income neighborhoods that are at risk, and transportation infrastructure, and virtually nothing's been done there because the total property tax proceeds is like 200,000 dollars.

Sometimes, these two understandings of "Who pays" are connected, with some interviewees saying that federal support is necessary to address SLR risk in low-income communities, for example. "Say if some little town that really doesn't have their money is affected [the legislature] needs to do something for the less affluent county, money has to start to come from somewhere" (interview 22).

The third most mentioned barrier is what we coded "vulnerability interdependence." Interviewees are aware of the interdependencies triggered by the geography of the Bay which render SLR a regional problem. We code this as a scientific barrier, but it does not refer to distrust in climate science. Rather, it refers to the complexity inherent in understanding which patterns of vulnerability interdependence exist across the Bay and the implications thereof for coordination between local jurisdictions. As interviewees 23 and 35 (respectively) stated:

> If we choose a spot, there's potential . . . and some studies show that actually if not done right, we're hurting other spots or other locations nearby. And in some instances, we need those jurisdictions or those jurisdictions have some leverage over the permitting of some of these projects, and so of course they're not going to want something that could adversely affect them. So maybe it's also scale, like the scale of the types of solutions that we can embark on would need to be pretty large.
>
> When a year or so ago, X was presenting his initial findings of some of his shoreline levee modeling, that a project here could have a response over there, a flood control manager for Santa Clara Valley Water District who's responsible for our projects within Santa Clara County, said, "Gee I guess we should be

talking to our counterparts in Alameda County," I mean, you know, anytime you do a project, what you do in the South Bay could have a response in the North Bay.

Awareness of vulnerability interdependencies across the Bay is important, because it signals the ongoing shift from the understanding to the planning phase of the climate adaptation planning cycle (Ekstrom & Moser 2014) and thus signifies progress has been made in the governance system.

The fourth most mentioned barrier is a classical challenge for all policies related to climate change (Bernauer 2013): the widespread perception that the problem will manifest itself far into the future discourages present investment given that those who pay today will not be there to enjoy the benefits tomorrow (Eisenack et al. 2014). Although acceptance that climate change is real is widespread in SF Bay (as shown in Figure 8 of Section 1), the scale of the investment required to future-proof the Bay to SLR discourages action, given the long time frames involved (according to current scenarios) before it manifests as a persistent issue. Interviewee 11 put it in simple words: "It's a slow burn, right. Sea level rise isn't like it's going to be here next week, it's not a hurricane kind of thing." We list this as an attitudinal barrier because interviewees related this concern to the attitudes of the general public of local communities in the Bay Area: SLR is too far off into the future for people to care ("The reason [dealing with SLR] is a hard thing to do is because people can't see it," interview 4) particularly as they struggle with other, more immediate and pressing challenges ("We hear a lot about flooding. A lot of the people don't really know why it's happening. We don't really hear about it because we have people being shot. There are other priorities that kind of come up" interview 16).

A dozen interviewees mentioned uncertainty as a barrier, in relation to both the timing of SLR and the planning choices required: what to plan for? How many centimeters or meters of SLR will actually occur at any given time? These barriers are typical of adaptation studies. For example, "conflicting timescales," "uncertainty," and "institutional fragmentation" are three of the barriers that tend to co-occur in the context of climate adaptation (Biesbroek et al. 2011). The funding barrier can be subsumed under the timescales problem as planning in the short-term for the long-term necessarily entails high upfront costs.

To these classic barriers, this study adds barriers that are peculiar (though not unique) to polycentric governance systems. For example, some of our interviewees lamented an information overload of SLR-related science. In a polycentric system, information is generated across multiple centers. While this encourages innovation (Ostrom 2010c), it can also result in over-dispersion, particularly when the pace at which the new information is generated is high. Some interviewees expressed the desire for information to be vetted, collated in one place, and communicated effectively in a way that is relevant to planning and/or policy-making. For instance:

> There are a number of big picture questions that need to be solved and solved over and over again in terms of for example having a regional data set of how do rising seas affect the edge of the entire Bay; obviously that data needs to get refreshed as new data comes in. Many planners are struggling with the fact that there are many, many data sets and it's hard to tell the set *de jure*. (Interview 34)

Further, we find that several of our interviewees mention regulatory barriers as an additional consequence of institutional fragmentation and call for more integrated permitting procedures for SLR projects; interviewees also mentioned current protocols for sediment management as problematic because they dispose of sediment instead of reusing it for SLR-related problems. Finally, and in line with much existing research on adaptation barriers, we coded several barriers related to political leadership: Notably, a third of our interviewees considered that politicians will not seriously intervene until a headline-grabbing disaster takes place in the Bay. Several lamented a general lack of vision for how to deal with California's exposure to the consequences of climate change in both the political establishment and the agencies' decision-makers.

Figure 11 shows counts of barriers by type. As per Ekstrom and Moser (2014), institutional barriers are the most numerous and most frequently mentioned, while scientific barriers are the least numerous. As mentioned, scientific uncertainty is mentioned as a barrier by fewer interviewees, suggesting a qualitative shift in their understanding from gathering information on the issue (typical of the understanding phase of the planning cycle) to attempting to devise a regional plan for the whole SF Bay (planning phase).

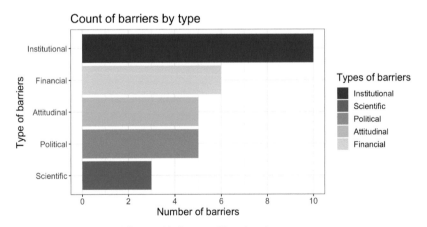

Figure 11 Count of barriers by type.

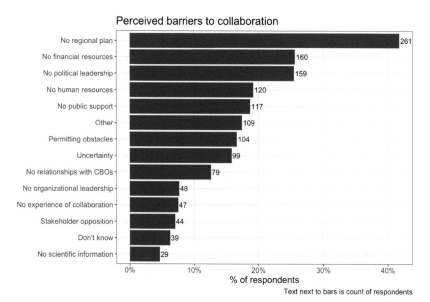

Figure 12 Survey respondents perceived barriers to collaboration.

Before we move to discussing solutions, we want to mention that in our online survey (fielded nearly two years after the interviews had taken place) we asked respondents to indicate the main barriers they perceive to engaging in collaborative activities related to adaptation to SLR in SF Bay. The results, reported in Figure 12, echo the findings of the interviews across the whole Bay: Respondents point to lack of a regional plan, political leadership, and funding as key barriers to their own engagement in the governance system.

Results: Solutions

We identified thirty-three solutions as emerging from the interviews. These are outlined in Figure 13. We color-coded them according to the same coding scheme used for the barriers. We deliberately coded the governance solutions so that most have an action, as indicated by a verb, in their title. The agent who is expected to carry out the action in the headings of the solutions is invariably either existing regional agencies or politicians.

Overall, interviewees mentioned more solutions than barriers (thirty-three solutions for twenty-nine barriers). Moreover, there is more overall agreement on the barriers (e.g., the most commonly mentioned barrier is shared by thirty-six out of thirty-nine interviews, while the most common solution is shared by twenty-seven interviews). This means that interviewees mostly recognized the same barriers but had different ideas on how to address them. This can be seen, for

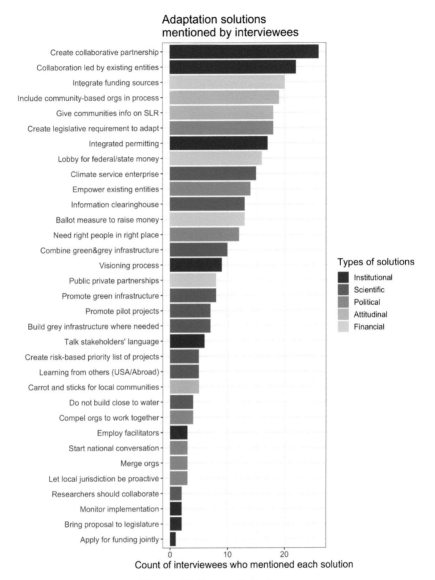

Figure 13 Count of solutions.

example, in Figure 13 by looking at the variety of scientific (and planning) solutions, with some interviewees advocating for "green infrastructure," some for "grey infrastructure," and some for a combination of both. At the same time, however, there is rather widespread agreement on a few core solutions (e.g., establishing a collaborative partnership).

As shown in Figure 13, twenty-seven interviewees mentioned "Collaborative partnership" as the chief solution they envisage for the Bay. Relatedly, most of them explicitly added that they would like the partnership to be led by existing regional agencies. This proposed solution underscores the familiarity of SF Bay with collaborative approaches that has been identified in previous research (Calanni et al. 2015). We depict these two answers separately to highlight the demand for leadership from SF Bay regional agencies that was explicitly expressed by our interviewees:

> Rather than start a whole new set of players and give them nothing to work with, I think it's better that the institutions who are here for good and know a lot more about it, are the ones that should band together and try to create a structure. (Interview 21)

> The impulse to want to reform government to make it more efficient and effective has often led to the creation of new government entities that sit on top of old government entities, and inevitably simply become yet another government institution that needs to be taken into account before you make a decision and move forward. (Interview 20)

Second, half of our interviewees remarked the necessity of integrating different funding pots at local and regional level into a larger SLR adaptation pot to finance SLR projects on the ground. Interestingly, many linked the necessity of local funding to climate change affecting different parts of the country in different ways, thus creating competing priorities for the federal government.

> Particularly when you look around the country there is a lot of need, so I don't know where the funding might come from. I don't think it's realistic to think local jurisdictions are going to be able to tax their way out of this. I don't think it's realistic to imagine that either the state or federal government is going to ride into the rescue. I think it's going to be pieced together. I have heard people talk of [Cap and Trade] money. That could be used. (Interview 14)

The following two solutions represent a two-pronged strategy to address the challenge of lacking community engagement: on the one hand, nearly half of our interviewees stated that community-based organizations have to be involved in the governance process, to take account of the vulnerability of those with lower resources; on the other hand, half of the interviewees stated that the general public needs to be presented with information on SLR in order to obtain public buy-in. Further, the distributional implications of planning for SLR do not go missing on our interviewees: Several expressed concerns about the inequality plaguing SF Bay and the vulnerability of communities who are both highly exposed and less able to cope. These same interviewees feared that more affluent communities would not agree to pool their resources to fund adaptation elsewhere in SF Bay – hence their insistence on making the concept of "interdependence" clear to the public.

The vulnerability interdependencies that exist in the Bay make SLR everybody's problem – this needs to be clear to local governments and communities (see first quotation below) but also to regional agencies, which, in a decentralized governance system like SF Bay, do not have powers over land use issues.

> We are going bottom up, and somewhere the bottom up and top down do meet. There is a need to think regionally. . . . But once the locals realize that this problem is so beyond their capacity, both, I think scientifically and engineering wise, and financially, that there will be an interest, and I think that will evolve. (Interview 37)

> The regional agencies don't manage land, so it's the land owners, the land manager, the coastal zone managers that need to come together because Association of Bay Area Governments is not going to build a levee, and Metropolitan Transportation Commission's not going to decide where a sea wall goes and Bay Area Air Quality Management District isn't going to decide where a horizontal levee is the right solution, it's going to be Foster City in San Francisco and County of Alameda in the City and Hayward and Marin County and the town of Belvedere, those are the people who are going to make the decisions. (Interview 1)

The solutions emerged from the interviews also speak to the peculiarities (and some of the disadvantages) of polycentric systems. For example, several of the solutions in Figure 12 ask political leaders to exert their authority to compel collaboration and planning where these are not undertaken voluntarily. Some interviewees would ask political leaders to create legislative requirements for developers to consider SLRs in their activities. Others would like policy-makers to compel municipalities to formulate adaptation plans or organizations to work together for climate adaptation if they refuse to take part in the collaborative partnership. A third of interviewees would also like political leaders to widen the powers of existing agencies and give them the formal authority to act where necessary. Further, over a third of the interviewees hoped in the creation of what we named a "climate service enterprise," that is, a set of people tasked with communicating the science and the interdependencies to both local communities and businesses and the public at large. In other words, our interviewees asked for a measure of centralized coordination. At the same time, however, Figure 5 shows that there is no appetite for new institutions to deal with SLR or adaptation as their sole task. Rather, there is appetite for governance and political leadership to foster a collaborative process between existing institutions (or mandate it via legislation if necessary).

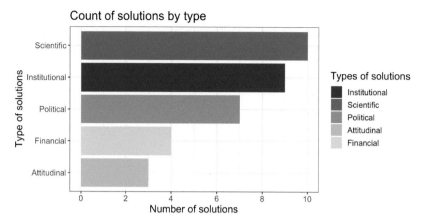

Figure 14 Count of solutions by type.

Figure 14 shows the counts of solutions by type. Most of the proposed solutions are coded as "scientific," that is, they concern technological or infrastructural solutions as well as learning from other jurisdictions. Each of these solutions has been mentioned by a small number of interviewees, suggesting different interviewees have different preferred technologies in mind to address SLR. There is much more convergence across interviewees concerning solutions to financial and attitudinal barriers.

Similarly to the previous section, we want to compare the findings of the interviews with the findings emerging from our later online survey, where we asked respondents to indicate up to three solutions that they think would speed up progress on adaptation to SLR in the Bay Area. The results, in Figure 15, report a similar picture to the interviews as concerns political leadership (with no appetite for new institutions) and collaborative partnerships. The main notable difference between the interviews and the survey is that a third of our survey respondents see investment in green infrastructure as a promising solution to SLR in the Bay Area, whereas our interviewees were more evenly split between supporting green and "grey" solutions.

Linking Barriers to Solutions

Finally, to show how barriers and solutions link together, we examined their co-occurrence across interviews in a matrix, independent of the category we assigned them to. The resulting heatmap is shown in Figure 16. For reasons of space and legibility of the figure, we keep only the barriers and solutions that co-occur in more than ten interviews. In the figure, darker colors indicate higher co-occurrence. The barriers and solutions do not have exact correspondence

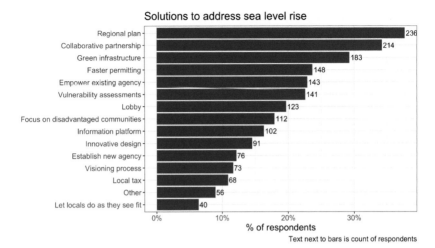

Figure 15 Survey respondents proposed solutions to SLR.

across the interviews (i.e., respondents sometimes mentioned solutions without the corresponding barriers, and barriers without solutions); the heatmap reflects this. Still, Figure 16 provides a coherent overview of the most salient topics which emerged in the interviews.

In Figure 16 we see that both barriers and solutions are found in all five categories: political, institutional, financial, attitudinal, and scientific. Yet, two institutional solutions, namely "create collaborative partnership" and "collaboration led by existing agencies" co-occur with all the most frequently mentioned barriers. These two institutional solutions co-occur in nearly all interviews with the barrier "jurisdictional fragmentation" but also co-occur very often with the barrier "who pays?" that is a financial barrier, with "vulnerability interdependence" which is a scientific barrier, and with "SLR too far off in the future" which is an attitudinal barrier. This shows that our interviewees see collaboration as a potential solution to most governance barriers, confirming the validity of the argument that collaboration is commonplace in SF Bay.

Discussion and Conclusion

We have several important findings from the analysis of the qualitative interviews, which shaped our understanding of the polycentric system and helped us set up the quantitative analyses presented in the next two sections. For one, we found confirmation of the familiarity of SF Bay with collaborative approaches to solving environmental policy problems. Second, we recognized a strong appetite for leadership in moving forward the adaptation cycle, while contending with issues of local control and the extreme fragmentation of the system.

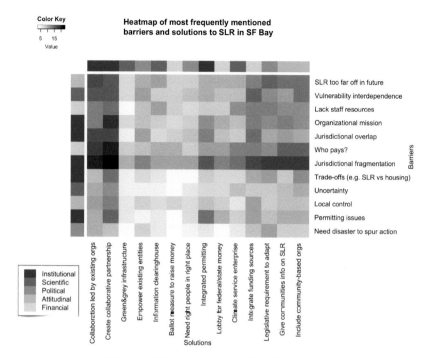

Figure 16 Barriers and solutions heatmap.

Third, we found that the barriers and solutions emerging from this sample of thirty-nine interviews and forty-two interviewees in 2016 are mirrored in those identified by the 722 respondents to the online survey of 2018. Fourth, we found that there is more agreement on the barriers than on the solutions. Finally, we found that our interviewees are aware of the psychological distance of SLR as a barrier to decisive policy action and widespread public support to fund-raising for adaptation.

In addition to these descriptive findings, we can draw some conclusions by relating our interviews to existing empirical findings in the literature. For example, in their study, Ekstrom and Moser (2014) predicted that the scientific challenges related to the inherent complexity of the vulnerability interdependencies between different locations of the Bay would become more salient in the planning phase. Awareness of vulnerability interdependencies comes through prominently in our interviews, suggesting that we observed the governance system for adaptation to SLR as it was concluding its understanding phase and entering the planning phase.

More generally, these findings force us to rethink the current understanding of the literature that thinks of adaptation as a private good (see Woodruff et al.

2020 for a similar call to rethink). That literature sees intervention, for example, to build protective infrastructure as benefiting only those who live in the location. We contend that adaptation is a public good, because it involves interdependencies between localities and across governance and geographic scales. Our interviewees understand these interdependencies; their existence is what fosters the demand for coordination and leadership that transpires so clearly from all interviews.

Moreover, the five categories of barriers and solutions are likely to be found in other contexts as well. Even in areas or countries where governance is on average more centralized, the cross-scalar nature of adaptation will inevitably lead to polycentricity because of the multiplicity of actors and sectors involved in planning for the resilience of affected locations. Hence, lessons drawn from SF Bay may apply in much different political systems.

Finally, the frequent appeals to "building trust" and "establish informal ties" that we collapsed under the category of "collaborative partnership" testify to this concept of leadership as a "device" to address the complexity of the adaptation task. In the next section, we operationalize this notion of leadership into our analysis.

4 Policy Networks for Cooperation and Learning

Introduction

To effectively respond to a new collective action problem like SLR, polycentric systems must mobilize networks of policy actors who participate across multiple policy forums to deliberate and collectively formulate adaptation plans and strategies. The key insight of much polycentric governance literature is that policy networks facilitate coordination by fostering the emergence of social capital that helps actors overcome collective action dilemmas (Berardo & Scholz 2010; Burt 2005; Lubell et al. 2014, 2017; Mewhirter et al. 2019). The policy network literature focuses on two key concepts related to the capacity of networks to respond to emerging collective action problems: social capital and leadership.

Social capital reflects that idea that the social relationships and processes inherent in social networks are productive in solving collective action problems linked to climate change (Adger 2003). The literature has focused on "bridging" social capital for coordination or "bonding" social capital for cooperation (Burt 2005; Coleman 1988; Granovetter 1973; Putnam et al. 1993). The words "coordination" and "cooperation" are borrowed from the jargon of game theory, while social capital is a concept derived from the literature of sociology. Thus, "coordination" refers to a situation (or game) in which actors want to achieve the same goal (in the famous "battle of the sexes" game, a couple want to go on a date but have different preferences on how to achieve it in practice);[2] once they coordinate their behaviors, deviating from the agreement is costly for all, thus there is no incentive to do so; this eliminates the need for monitoring each other's behavior. In contrast, "cooperation" refers to a situation (or game) in which actors have a common goal, but one of them can benefit from deviating from the agreed-upon strategy (as in the prisoner's dilemma game) or can free-ride on the efforts of the others. The presence of these incentives to deviate from the agreed-upon strategy requires actors to monitor each other's behavior.

The policy network literature transferred these ideas into the context of policy networks by hypothesizing that actors collaborate to achieve either coordination or (to overcome the challenges of) cooperation (Berardo & Scholz 2010). In this thinking, the presence of so-called "open" network structures constitutes "bridging social capital" that actors use to address coordination problems. Open networks are centralized around key policy actors who bridge geographic and group boundaries, as actors seek to learn about the preferences and motivations of other participants to find the coordination equilibrium. Open networks are useful to this task because they efficiently diffuse information and thus facilitate policy learning. In contrast, the presence of so-called "closed" network structures constitutes "bonding social capital" that actors use to address cooperation problems. "Closed" structures consist of reciprocal connections and transitive, friends-of-friends type relationships (e.g., triangles, where if actor i is connected to j and k, actors j and k also share a tie) that actors use to enable the maintenance of trust, norms of reciprocity, and sanctioning mechanisms.

As for leadership, empirical research finds that policy networks feature a "core" of policy actors who provide connectivity to the rest of the network (Angst & Brandenberger 2022; Angst et al. 2018; Berardo & Scholz, 2010; Lubell et al. 2016; Vantaggiato & Lubell, 2023; Yi, 2018). This core is obvious during participatory research – the same organizations and individuals are repeat players across multiple forums where they develop trust and know each other as individuals. In existing research (Vantaggiato & Lubell 2023), we call these core actors "leaders" and contend that they support both learning (for coordination) and cooperation. For learning, leaders are well-informed about a collective action problem like SLR; other actors consider them as experts with legitimate information; for cooperation, leaders reduce transaction costs across the network and persuade other members of the network to engage in coordinated behavior. This

---

[2] Famously, the husband preferred to go see ballet at the theatre while the wife preferred to see a football match in a bar.

double role translates into a coexistence of network configurations (or "motifs," see Bodin 2017) related to coordination and cooperation. We theorized this coexistence as representing "multifunctional social capital" supporting multiple social processes (learning, cooperation, bargaining) at the same time.

In this section, we apply what we know about social capital and leadership in policy networks to understand the structure of the policy network for SLR adaptation in SF Bay as a polycentric system. Our qualitative analysis in Sections 2 and 3 suggested that the polycentric system of SF Bay was moving from the understanding to the planning phase of the adaptation planning cycle. In this section, we contend that leaders play a role in facilitating this transition (Vignola et al. 2017) and that we can see this in the motifs that structure the network, once the action of leaders is taken into account. In what follows, we introduced the theoretical framework and test this hypothesis using a Bayesian Exponential Random Graph Model (ERGM).

Theory: Social Capital and Leadership

This section provides the background for our two arguments about multifunctional social networks supported by leadership in the core of the network. The two arguments combined suggest that observed networks will have structural motifs that reflect bridging and bonding social capital and are hypothesized to support learning and cooperation, respectively. The networks will also display core–periphery structure where leaders can facilitate the evolution of the polycentric system to solve new collective action problems and other actors form bridging ties looking for information and bonding ties based on recognized interdependencies and resource needs.

Multifunctional Social Capital

The multifunctional social capital hypothesis is rooted in the earlier "risk hypothesis," which posits that policy networks are structured to provide bonding social capital that is useful for addressing cooperation problems (where the risk of collaborators' defection is high), or bridging social capital that is useful for addressing coordination and information-sharing problems where the risk of collaborators' defection is low and common knowledge is needed to orchestrate decisions (Berardo & Scholz 2010; Burt 2005). Because solving coordination problems requires developing and sharing information, the process can be considered a type of policy learning (Heikkila & Gerlak 2019).

Each type of social capital is thought to correspond to specific network motifs and social processes (see Table 2): sets of microlevel social interactions that coalesce to form a policy network. Bridging social capital corresponds to "open" network motifs (so-called "open two paths" and "network centralization," that is, the existence of highly central actors in the network), which actors use to reach knowledgeable network members as directly as possible. Bonding social capital corresponds to "closed" network configurations or "motifs" (reciprocal ties and triangles, i.e., friend-of-a-friend relationships) that actors use to share information with each other and monitor each other's contribution to the common effort.

The fundamental insight of Berardo & Scholz (2010) was the idea that one can deduce the type of collective action problems policy actors are grappling with by analyzing the structure of the network emerging from their collaborative ties with each other: if the network is centralized, bridging social capital predominates which means actors deal with coordination problems; if the network is decentralized and rich in triangles, then bonding social capital predominates which means actors deal with cooperation problems. The empirical strategy is then to observe

Table 2 Social capital network motifs.

Function		Network Motif
Coordination Structures (Bridging social capital)		
Degree Centralization	Leadership and brokerage	
Open Two Paths	Information transmission	
Cooperation Structures (Bonding social capital)		
Reciprocity	Trust-building and monitoring of the cooperation effort	
Transitivity	Trust-building and monitoring of the cooperation effort	

the network structures and decide whether they are responding to the type of collective action problem for which they are well-suited.

In contrast, the multifunctional hypothesis is derived from the EGF and argues that solving new collective action problems requires multiple social processes of learning, cooperation, and bargaining (Levy & Lubell 2017; Vantaggiato & Lubell 2023). The actors must learn about the causes, consequences, and possible solutions to the collective action problem as well as the preferences of other actors. They must cooperate in the context of developing plans and implementing strategies. Because the potential set of policy solutions features heterogenous distributions of cost and benefits, they have different policy preferences and bargain over potential agreements. Thus, policy networks in polycentric systems need to provide structures capable of supporting multiple social processes at the same time, particularly when dealing with new, emerging collective action problems for which no blueprint of behavior and policy solutions exists. Thus, rather than expecting one type of network motif to dominate depending on the type of collective action problem, we expect coordination and cooperation motifs to coexist. Indeed, even in the initial empirical analysis of the risk hypothesis, there is evidence that both families of network structures operate at the same time (Berardo & Scholz 2010; Yi 2018).

Leadership: Fostering Coordination in Policy Networks

Network structures arise from the aggregation of microlevel interactions (what we called "motifs" in Table 2) between individuals in their environment (Desmarais & Cranmer 2012). Typically, actors/organizations found in the core of a policy network have more ties, on average, than most other actors.[3] The findings of empirical research are that core actors are typically governmental actors, who have the incentives, authority, and capacity to lead the network toward collective goals, while peripheral actors are typically less resourced, for example, nongovernmental organizations and local governments, who respond to information and resources provided by the core (Angst et al. 2018; Bodin & Crona 2009). Thus, actors in the core are typically referred to as "coordinators" or "brokers" (Angst & Brandenberger 2022; Burt et al. 2021; Lubell 2004) or "leaders" (Burt et al. 2021). This understanding of core position as leadership comports with the notion that leadership is relational (Ahlquist & Levi 2011) and therefore has structural implications, which network models should be able to detect.

---

[3] Possessing many ties is not a guarantee of core-ness: an actor could have many ties to organizations which are, themselves, not very well connected and end up in the periphery of the network. Thus, core actors possess more ties on average, and more ties to other well-connected actors.

In this section, we build upon our existing research, where we used community detection to identify leaders and followers in the policy network of adaptation to SLR in SF Bay and conceive the multifunctional hypothesis (Vantaggiato & Lubell 2023). The task of community detection was complicated by the fact that the policy network of adaptation to SLR in SF Bay has core–periphery structure. In a core–periphery network, the core is where most of the clustering (i.e., the triangular friends-of-friends relationships) occurs; this contrasts with a typically much sparser periphery, where there typically are fewer triangles and fewer ties between actors. This structure makes the community detection task more challenging (Yang & Leskovec 2012).

To perform community detection on the core–periphery network of SF Bay, we used the so-called Affiliation Graph Model (AGM) (Yang & Leskovec 2012) – a community detection algorithm for core–periphery networks. The AGM deals with the imbalance between the high density of the core and the sparsity of the periphery by clustering together nodes that have similar patterns of connections, i.e., are affiliated with the same social circles. In Vantaggiato & Lubell (2023), we used the AGM to identify the cluster of actors who belong to all communities in the SLR policy network. Empirically, we retrieved five communities and twelve leading organizations for the network of 612 organizations we observed. The leadership group comprises governmental agencies (state, regional, county, and local) as well as universities and research centers. In essence, these actors are interchangeable; they connect all the communities and are fully connected with each other. They are listed in Table 3. We will use "leadership" as an individual attribute in the statistical analysis.

Hypothesis: Social Capital and Leadership as Engines of Transition between Phases of the Adaptation Planning Cycle

By combining the insights of the literature on policy networks and on the adaptation planning cycle, we can conceive of a new hypothesis to identify the phase of adaptation a polycentric system is in. Namely, observing predominantly bridging motifs (i.e., a very centralized network with few triangles and low reciprocity) in a policy network of actors dealing with adaptation would suggest that the system is still wholly in the understanding phase: the main EGF social process is learning, and actors search the network for information on SLR and their own vulnerabilities. They seek coordination, being aware that the problem has a regional dimension and that no one actor can address it on their own. They form open network structures which do not require monitoring of others' efforts and allow reaching well-connected others.

Table 3 List of network leaders in SF Bay.

Name of the organization	Description of the organization
[1] "Bay Area Regional Collaborative (BARC)"	State agency established to coordinate planning and implantation around transportation, air quality, local governments, and coastal management.
[2] "Bay Conservation and Development Commission (BCDC)"	State agency with regional jurisdiction, responsible for permitting coastal development in SF Bay and planning for coastal issues.
[3] "California State Coastal Conservancy"	State agency that focuses on protecting coastal land-uses and providing access.
[4] "Federal Emergency Management Agency (FEMA)"	Federal agency that is responsible for flood zone and other emergency management.
[5] "Marin County"	North SF Bay County that is innovator in climate adaptation, and includes active environmental conservation and environmental justice groups.
[6] "National Oceanic and Atmospheric Administration (NOAA)"	Federal agency focused mainly on ocean management and weather.
[7] "Port of San Francisco"	Special district that provides international critical transportation and shipping infrastructure.
[8] "Resilient by Design"	High profile design competition for developing innovative sea level rise adaptation concepts.
[9] "San Francisco Bay Regional Water Quality Control Board"	State agency responsible for implementing water quality regulations.
[10] "San Francisco Public Utilities Commission"	Local special district that provides drinking water and wastewater services to the city of San Francisco
[11] "United States Geological Survey (USGS)"	Federal agency that is responsible for producing the science that is linked to policy decisions, and has invested heavily in sea level rise research.
[12] "University of California Berkeley"	University of California campus that has leading researchers developing climate change and infrastructure models.

In contrast, if we observed predominantly bonding motifs (a decentralized network, rich in reciprocal ties and triangles), we would conclude that the system is past the understanding phase and into the planning or even the implementation phase where, as we posited in Section 1, conflict may be rife as different planning options entail different distributions of costs and benefits for different regions of the Bay and different groups. At that stage, actors may have formed partnerships or coalitions of interest (Berardo & Scholz 2010), lobbying for their own side of SF Bay to be protected first. This would correspond to the EGF social process of building cooperation within coalitions (Schlager 1995) and, therefore, would require bonding social capital for actors to monitor each other's behavior and maintain close communication.

If instead we observed both types of social capital (a network that is both highly centralized and rich in reciprocal ties and triangles) in the structure of the network of relationships between SF Bay policy actors, this would suggest that the polycentric system is still in a transition phase between understanding and planning and thus displays both types of motifs, as per the multifunctional hypothesis (Vantaggiato & Lubell 2023). Highly central leaders provide the bridges for local actors to both seek coordination and develop cooperative arrangements at local level.

Our analysis of the interviews suggested that, at the time of our observation, the polycentric system of SF Bay was transitioning from the understanding to the planning phase: while institutional fragmentation and funding remained key concerns, awareness of vulnerability interdependencies and proactive thinking of different funding portfolios suggested the system was narrowing down on planning options. As mentioned in Section 3, however, we fielded our online survey two years after the interviews – the system might have moved further along the adaptation planning cycle. We will make these assessments via examination of the results of a Bayesian Exponential Random Graph Model (BERGM). ERGMs are generative models of network structure, i.e., the assumption is that the network evolved into its current structure and that the model helps uncover the drivers of its evolution.

Data Collection

We adopted a data collection strategy inspired by adaptive design (Handcock & Gile 2007), which exploits the components of the network as they are observed to guide sampling, coupled with triangulation between different sources of data (interviews, survey, policy reports). As explained in Section 2, we disseminated the survey in summer 2018 and had a response rate of 23 percent in terms of individuals, for a total 722 respondents who completed or partially completed all sections of the survey.

The survey asked respondents the following question: "please list the organizations that you have collaborated most closely with in the context of sea level rise in the past year" for each of the following types of actors: federal, state, regional, local, and nongovernmental actors. Respondents could list up to twenty actors for each category. This name-generator approach presents some shortcomings in that it tends to measure fewer linkages among a broader set of actors (Henry et al. 2012). We decided to adopt it given the sheer number of governance actors involved in sea level rise in SF Bay, which promised to render the roster approach (where survey respondents are asked to check their closest collaborators from a list) unwieldy and overwhelming for respondents.

A total of 443 individual respondents replied to the question. Of these, forty-four declared being involved in an individual capacity (these comprise consultants, academics, and retired local officials), while 399 declared that they responded on behalf of an organization. These 399 respondents collectively represent 256 organizations. When more than one respondent belonged to the same organization, we merged their replies, considering that multiple respondents from the same organization suggest higher involvement of that organization in the governance system. Survey respondents named a total of 464 organizations as collaborators. Of these, 180 are also among the total 878 respondents to the survey, but only 153 of them answered the network question, while 284 are not among the respondents. After cleaning the data and excluding isolates (i.e., actors with no ties to other actors), the network we analyze comprises 612 actors.[4]

We consider that our sample of actors is representative in the sense that the average number of relations in the inferred network that an actor with a particular value of a given attribute (e.g., governance level) has with other actors with a particular value of a given attribute (in this case, the same attribute) is close to that of the sample and thus that of the population. In other words, we consider that the sampled data has a degree distribution similar to that of the population and similar mixing properties (Krivitsky et al. 2011). Therefore, we

---

[4] The data cleaning process of the network data has been painstaking, and it has involved making many coding decisions. That process has been replicated by two different researchers within the research group of one of the authors, leading to different numbers of actors (+ or – 30) included in the network depending on the coder's decisions concerning how to consider respondents who did not fill in the network question of the survey and those who responded as individuals. We excluded the former and included the latter. We excluded isolates from the Bergm model to prevent the indegree variation from being overstated; we included individual respondents in order to keep as many of the named collaborators as possible. We replicated the analysis in this section with the network data obtained via this different coding procedure: the results are unchanged bar for the indegree term, which is significant and negative in those analyses.

proceed under the assumption that the network boundary is well specified and that our missing data is ignorable (Handcock & Gile 2007), that is, that there are no systematic differences between respondents and nonrespondents (Little & Rubin 2019).

Methods

Bayesian Exponential Random Graph Model (BERGM)

We test our hypotheses linking social capital to social processes and phases of the adaptation planning cycle by using Bayesian Exponential Random Graph Models (BERGMs). A simple explanation of what ERGMs do is that they count microlevel configurations (Desmarais & Cranmer 2012) in the observed network and compare those counts to the counts that would occur in a network with similar characteristics (e.g., number of nodes and density) generated at random. In their turn, BERGMs (Caimo & Friel 2011) allow for the inclusion of prior information about the data generating process and/or from previous research through an informative prior distribution. This can be done by placing prior probability distributions on the possible values of the unknown parameters or models. Thus, one can incorporate into the ERGM formula the prior means of the values of the structural motifs that are typically included in such models, for example, reciprocity and transitivity for bonding social capital, degree centralization and open two-paths for bridging social capital (see Table 3). Including priors allows researchers to create a "null model" to test their network against and see whether its structure conforms to those priors or not.

Analysis: One-Mode Policy Networks

Descriptives

Figure 17 depicts the policy network for SLR adaptation in SF Bay, with nodes sized by degree and colored by the type of organization they represent. We collapse federal, state, and regional actors into the same category of actor ("regional govt" in Figure 17) for the sake of the legibility of the graph. As can be seen in Figure 17, regional government actors are at the core of the graph and have the highest number of connections (the nodes in the graph are sized by their degree, i.e., their number of connections to other nodes). However, leadership does not necessarily equate degree centrality; some of the highest-degree nodes in our network are not leaders. What makes for leadership is diversity of connections across multiple functional communities (Vantaggiato & Lubell 2023). In this section we take the leaders as given and ask: Once the leaders are accounted for, what type of social capital structures this policy network?

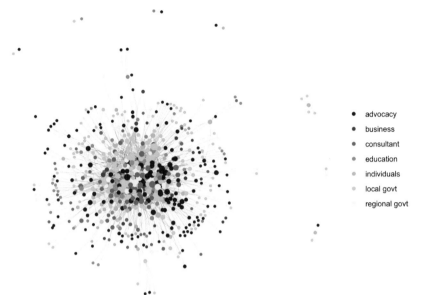

Figure 17 The network of SLR adaptation governance in the San Francisco Bay Area, nodes sized by degree.

To make sure that we are capturing the social processes we think we are capturing, our survey asked respondents to indicate the importance they assign to different criteria for partner selection in their collaborative activities concerning SLR in the Bay Area, on a scale from 0 to 10. As shown in Figure 18, our respondents assessed reputation as the overall most important attribute of prospective collaboration partners, closely followed by possession of information and the ability of affect one's interests as well as the possession of actual decision-making authority. Thus, most respondents want to partner with reputable, reliable, trustworthy collaborators. Respondents have a strong preference for partners that "affect one's interests" indicating that they understand that they are interdependent in the achievement of adaptation. We expect these preferences to lead respondents to form reciprocal relationships and to close many triangles, that is, the structural fingerprints of bonding social capital, which fosters the emergence of trust. At the same time, respondents choose partners with information that is useful to the process, indicating that learning is (still) an important process directing network ties. Partners who do different activities than oneself have a broad network of contacts and possess ample resources that complement this set of criteria and predict ample bridging social capital in the network. Choosing partners who carry out the same activities is the least important criterion for

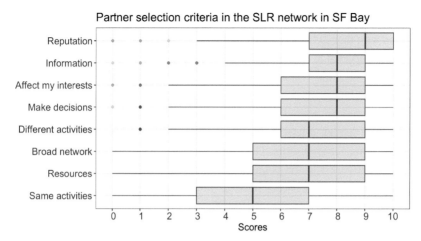

Figure 18 Partner selection criteria in SLR policy network.

partner selection, bespeaking awareness of needing to collaborate with different actors. Taken together, these descriptive data suggest support for the multifunctional hypothesis.

Bayesian ERGM Results

We modelled the network as directed. This allowed us to test for the reciprocity term of bonding social capital and for the popularity of leaders and other actors as measured by indegree (i.e., the number of ties an actor receives from others). However, our data comprise many actors who were named as collaborators but did not answer our survey. To account for this fact, the model includes a matrix of structural zeros in outdegree for those actors. This prevents the density term from being overstated and provides for a more accurate estimation of the other terms. We also included a binary covariate marking the 12 leaders identified in Vantaggiato & Lubell (2023) as 1 and all other actors as 0.

The priors set for the model include a negative density (i.e., the average probability of a tie in the network is 0.01), positive reciprocity and transitivity, a weak negative prior for indegree (which would indicate centralization on influential actors, a common feat of bridging social capital), and a zero effect for homophily (i.e., the tendency to form ties with actors similar to oneself in some attribute, see (McPherson et al. 2001)) by type of actor based on the findings in Figure 18. Finally, we set a high prior for the coefficient on leaders' popularity (i.e., their tendency to receive ties), as they are the only actors belonging to all the functional communities of SLR in the Bay.

Figure 19 shows the results of the model by plotting the mean parameter values with their 95 percent credible interval. The goodness-of-fit plot for the model is reported in the Appendix (Figure A1).

To interpret the results of the model, one must bear in mind that, in ERGMs, coefficients represent the change in the log-odds of a tie for a unit change in a predictor. Similar to the interpretation of a standard regression model, a new tie added to the network changes the probability of observing certain motifs. The effect of adding a new tie to the network on a given network statistic is modelled using the "change statistic," which is the difference in the network statistic (reciprocity, triangulation, homophily, etc.) before and after adding the tie. To calculate predictions, one must use the summation of the model coefficient estimates multiplied by the change statistics for each variable as the quantity to be exponentiated (Ready & Power 2018).

For example, the tie probability at the intercept (i.e., the density parameter) for the cooperative network is $\exp(-5.79)/(1 + \exp(-5.79)) = 0.003$ when all other terms are zero. This is lower than the actual density of the network, which is 0.01, because it does not take into account the reciprocity, the homophily, the leadership, and so on. Further, the coefficient for indegree is not significantly different from zero; which means that variation in indegree is low. In other words, once the leaders are taken into account, the transitivity in the network (often centered on leaders) predominates over centralization (i.e., the tendency of networks to be structured around popular actors). As expected, reciprocity and transitivity are positive, as is the tendency of

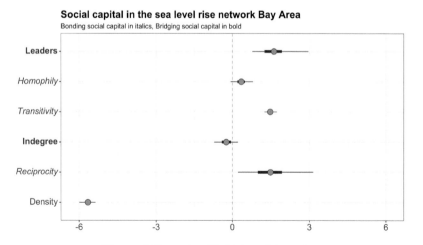

Figure 19 Bayesian ERGM model results.

leaders to receive ties, while homophily is not significant (meaning that actors are not more significantly likely to establish ties with actors of their same type). Recall that "leadership" is not the same as "centrality": we identified leaders using a community detection algorithm for core–periphery networks (Yang & Leskovec 2012). There are nodes in the network with higher indegree than some of the leaders. Yet, the leaders are clearly very central actors in the network.

To understand what the model means, consider that the probability that any new tie added to this network is directed to one of the leading organizations and does not create any new triangles or reciprocate any existing tie is very low: the log-odds of the density term plus the leadership term amount to $-5.79 + 1.5 = -4.29$, which gives $\exp(-4.29)/(1 + \exp(-4.29)) = 0.01$. Instead, an hypothetical new tie is much more likely to, at the same time, be directed to one of the leading organizations, being reciprocal and closing a triangle – a probability of 0.80 (Ready & Power 2018). The probability that a new tie added to this network is reciprocal and forms a triangle but does not involve any of the leaders is still high, but much lower at 0.52. These figures mean that the leaders play an enormously important connecting function that facilitates the emergence of bonding social capital (transitive ties and reciprocal ties). Taken together, these results indicate that leaders provide a strong foundation of both bridging and bonding social capital to this network while most other actors interact at a smaller scale, fostering reciprocal and transitive bonding relationships as they begin addressing the cooperation problems that we expect to be associated with the planning phase of the adaptation planning cycle.

Discussion and Conclusions

In this section, we posited a link between social capital and leadership, network motifs and the adaptation planning cycle. Namely, we argued that observing a predominance of bridging social capital would indicate that coordination/learning is the main social process ongoing in the system, which in turn would mean the system is in the understanding phase of the planning cycle. In contrast, observing a predominance of bonding social capital would lead us to argue that actors are trying to build cooperation, suggesting the system had reached the planning phase. Alternatively, the system may still be in his transition from the less controversial stage of understanding (and learning about SLR and SF Bay vulnerabilities) to the more controversial stage of planning (i.e., deciding what options will be pursued), and we would observe both types of motifs as constituting the network.

Importantly, we expected leadership to emerge as the engine of this transition from understanding to planning as leaders facilitate both exchange of information and the emergence of cooperation and coalitions. The results of the BERGM model support this third conjecture: bonding and bridging social capital coexist in the network, suggesting both processes of learning and cooperation (in EGF parlance) are ongoing, that the network was still in transition to the planning phase, and that leadership (exemplified by twelve leading organizations identified in previous analyses) facilitated the system moving forward (Vignola et al. 2017).

Yet not all the action takes place in the core. While the leaders-group comprises mostly governmental actors and knowledge providers, the periphery is rich in local governments, special districts, and local community-based organizations. This does not imply that local actors are "passive" receivers of the instructions of the leadership core. To the contrary, we know that in California and in the Bay Area "Local is King." Governance is highly decentralized and fragmented, with local governments and special districts wanting to maintain their autonomy and rejecting any form of "control" from higher levels of government (Ostrom et al. 1961). We surmise that these actors look to the core for information but maintain bonding relationships at local level that do not necessarily involve the leaders.

To accomplish their connecting tasks, leaders set up policy forums to attempt to solve the collective action problems that occur in the polycentric system. In the next section, we examine several "policy games" (in EGF parlance) chosen because of their visibility and importance in the governance system and assess the predictors of participants' perceptions of their performance.

5 Sea Level Rise Adaptation Games
Introduction

In the previous section, we argued that leaders of multifunctional policy networks catalyze how a polycentric system moves through different stages of the adaptation planning cycle. Another key feature is institutional change, which occurs both through change in existing policy forums and through the creation of new policy forums explicitly devoted to an emerging collective action problem. This section will analyze the features of seven policy forums that were created to address SLR in SF Bay and how actors perceive these forums' performance for learning, cooperation, and tangible progress in addressing SLR.

In other research (Lubell et al. 2020; Vantaggiato & Lubell 2022), we have studied the determinants of forum performance in polycentric systems by focusing on learning and specialized knowledge within the whole ecology of policy games for which we could collect data. In contrast, this section focuses on what we call the "big games in town." The "big games" attract a high level of participation and therefore are the center of the networks embedded in polycentric systems. Just like the Southern women desiring to attend the big parties (Davis et al. 1941) policy actors frequently discuss the big games and do not want to miss out on participating.

Moreover, polycentric systems typically comprise different types of forums, devoted to different purposes ranging from information diffusion to planning to implementation. The polycentric governance system is multifunctional not only in the policy networks that emerge from actors' interactions, but also in the kinds of forums that exist. Studying actors' perceptions of policy forums with different purposes provides yet another angle for us to gauge how barriers to adaptation interact with the features of a polycentric governance system. Analyzing the big games also provides empirical benefits because there are enough participants to measure the average level of performance. From a qualitative perspective, we also know enough information and the specific details of each game, to provide some process-based insights on the observed results.

Of theoretical concern in this section is variation in how different types of participants perceive the performance of the forums they attend based on their role in the system and the purpose of the forum. The literature finds that some actors (typically government agencies) play a "neutral" role trying to foster coordination between other actors (Angst et al. 2018; Lubell 2004) or serving as brokers of information between the center and the periphery of the network (Vantaggiato & Lubell 2023) or across political divides (Angst & Brandenberger 2022). In contrast, other actors play an advocacy role around specific issues, advocating for the interests of their members or constituency. This includes nongovernmental organizations as well as certain types of local government such as special districts. These actors are "partial" toward certain policy solutions and participate in the governance process to advocate for their specific policy preferences.

Across both types of actors, some specialize in sea level rise and climate adaptation and thus have a vested interest in creating and sustaining related policy forums. Other actors approach SLR as an issue that intersects with their other policy interests or responsibilities. For example, water districts worry about SLR because it may affect the capacity of wastewater or drinking water infrastructure. Environmental groups worry about SLR because it may affect ecosystem function or species habitat, while community groups are concerned about how SLR may affect housing availability. The extent to which a particular actor specializes in SLR is thus another individual factor that will affect their perceptions of policy forums.

In the following, we explain why the "big games" are worth studying and why we can expect perceptions of forum performance to vary based on our classification of actor types into "neutral" (governmental actors and researchers) and "partial" (advocacy groups). Using data from our 2018 online governance survey, we run multilevel regression models which show that "partial" actors are more likely to assign lower scores to forums' performance than "neutral" actors particularly in terms of outcomes (impact and effectiveness). This suggests that the governance system does not reflect the interests of (at least some) advocacy actors. Our interpretation of this finding is that the "big games" are still too focused on addressing the barriers typical of the understanding phase of the planning cycle; "partial" policy actors, however, may be looking for forums where they can debate and defend their preferred planning options. This may be an additional reason why the polycentric governance system for adaptation to SLR in SF Bay struggles to move forward: the leaders (who specialize in convening stakeholders and providing knowledge) and the policy forums they promote are not in sync with the demands of the followers. This may be because the leaders do not possess the authority or the legitimacy to deal with the political and distributional conflicts inherent in selecting planning options, strategies, and on-the-ground projects.

Big Games in Polycentric Systems: Like Bees to Honey

Polycentric systems often include hundreds of policy forums in which actors participate and deliberate over specific issues. The EGF defines policy games as constellation of actors, issues, and venues (or forums), because this constellation defines who is playing the game, their preferences, and payoffs for different strategies. This is an explicitly game-theoretic conceptualization of polycentric systems, although it is very difficult to precisely measure the payoffs available in each game. We expect the payoffs to vary across each game, and we know empirically that different games experience different levels of conflict and cooperation (Lubell et al. 2020).

While theoretically the policy process unfolds across all the games in a system, there are always some "big games in town" that attract more participation and visibility than other games. Policy actors are drawn to these big games "like bees to honey," because they offer more policy resources, legitimacy, political authority, and visibility. Analogous to the network process of "preferential attachment" (Barabási 2009), if powerful actors seek to participate in the same policy games as other policy actors, nobody wants to be "left out of the party." Hence the "big games" will typically have more influence on the overall governance system. The learning, cooperation, and bargaining that happens in big games will spread through the system as actors participate across games, and institutional change and resource decisions made in the big games will structure payoffs in more peripheral games.

Forum Type and Perceptions of Forum Performance

Policy forums do not prominently feature in the literature on the adaptation planning cycle, which tends to focus on individuals and their perceptions of adaptation barriers rather than their perceptions of the venues where they discuss them. We contend, however, that actors realize the barriers and governance challenges they face from the interactions they have within policy forums. The clash of preferences and priorities may emerge from within the process of trying to agree on potential solutions. For this reason, policy forums are as integral a part of the adaptation planning cycle as policy networks.

The literature on polycentric governance systems provides important insight on policy forums. Fischer and Leifeld (2015) put forward criteria to systematically map and assess policy forums. They point to structural features such as size of the forum and to process-related features such as rules in use and mechanisms of inclusion of all relevant actors. Maag and Fischer (2018) underscore that different forums have different purposes and participants pick the forums they attend based on their mandates and goals. Fischer and Maag (2019) find that actors attribute different importance to the forums they attend and appear to privilege forums that lead to learning and resource distribution.

Our analysis identified seven big games[5] at three levels of geographic scope: local (counties or below), subregional, and regional. At the local level, Marin BayWave is a vulnerability assessment of Marin County's eastern shoreline which began in 2015 and was released in 2017. The work of BayWave continued until 2020 when Marin released its Adaptation Planning Guidance.[6] The San Mateo County "SeaChange" initiative (launched in 2015) is another example of a local vulnerability assessment, released in 2018.[7] Both BayWave and SeaChange were sponsored by county governments and featured ample community engagement.

At the subregional level, the State Road 37 (SR 37) Policy Committee[8] (formed in 2015 and still active) focuses on developing adaptation alternatives for a low-elevation, major roadway that crosses multiple counties in the North Bay. At the regional level, BayCAN[9] is a collaborative partnership formed in

---

[5] We classify as a "big game" any forum mentioned by over twenty survey respondents.
[6] https://www.marincounty.org/main/sea-level-rise/baywave (last accessed 13 August 2023).
[7] https://seachangesmc.org/vulnerability-assessment/ (last accessed 13 August 2023).
[8] https://www.tam.ca.gov/sr-37-policy-committee/ (last accessed 13 August 2023).
[9] https://www.baycanadapt.org/what-we-do (last accessed 13 August 2023).

2018 and designed to coordinate and inform local governments, while the Coastal Hazards Adaptation Resiliency Group[10] (CHARG) is an initiative launched in 2014 by the Bay Area Flood Protection Agency Association (BAFPAA) that develops networks around special districts. Both BayCAN and CHARG are still active. An early SLR program spearheaded by the Bay Conservation and Development Commission in 2010, Adapting to Rising Tides[11] (ART) supported vulnerability analysis at the local level and also developed a Bay-wide vulnerability analysis. Resilient by Design[12] (RBD) was a one year (2017–18) regional planning competition that solicited innovative adaptation projects across SF Bay. RBD was modeled after the Rebuild by Design competition held after Hurricane Sandy in the New York/New Jersey region. While these policy forums do operate at a specific geographic scope, it is important to note that they interact with other policy forums and policy actors at both higher and lower levels of geographic scope. Concomitantly, many of the same organizations and individuals participate in overlapping forums, and thus interactions within the forums are a catalyst for developing the policy networks examined in the previous section.

Importantly, the big games also evolve over time. For example, CHARG was quiescent for some time and then revitalized under the auspices of BAFPAA. Furthermore, the quantitative analysis considered here does not include the emergence of new "big games" such as BayAdapt, which was the main regional SLR planning process starting in 2020 and continues to be a focal point for implementation. Indeed, it is fair to say that BayAdapt would not exist without the learning, cooperation, and bargaining that occurred during the overall evolution of the system. The dynamics of policy forums – their birth, change, and (rarely) death – is a key feature of polycentric systems, and high-capacity systems like SF Bay feature a dynamic ecology of games.

In our research, we found that planning forums harbor the most conflict in the context of adaptation to SLR in SF Bay (Vantaggiato & Lubell 2022), while forums focused on exchange of information and networking, scientific dissemination and vulnerability assessment harbored the least. Similarly, Lubell et al. (2020) find that organizational heterogeneity and number of actors is more highly associated with conflict than the diversity of issues considered in a forum. This empirical evidence suggests that conflict arises when actors with diverse resources, mandates, and preferences deliberate about options for on-the-ground projects (i.e; the planning phase of the adaptation planning cycle)

---

[10] https://sfbaycharg.org/who-we-are/ (last accessed 13 August 2023).
[11] https://www.adaptingtorisingtides.org/about/ (last accessed 13 August 2023).
[12] http://www.resilientbayarea.org/about (last accessed 13 August 2023).

rather than when actors attend forums to access information and learn about the issue (i.e; the understanding phase). This pattern suggests that policy forums may mirror the phases of the adaptation planning cycle (understanding-planning-implementation) and thereby also the types of barriers and conflicts inherent in each.

These earlier results lead us to formulate the following hypothesis:

H1: The performance of big games is highest for forums focusing on the "understanding" task of the climate adaptation planning cycle.

Actor Type and Perceptions of Forum Performance

We are interested in examining whether different policy actors perceive the performance of the policy forums they attend differently, depending on their role in the system. To operationalize actors' roles, we categorize respondents according to a binary attribute reflecting their stance in the governance system of SLR: "neutral" and "partial." Neutral actors are usually governmental actors, who typically adopt a neutral or brokering strategy that attempts to find compromise among competing interests, as evidenced by their frequent sponsorship of policy forums (Angst & Brandenberger 2022; Angst et al. 2018; Ekstrom & Moser 2014; Hamilton & Lubell 2017; Lubell et al. 2014; Meijerink & Stiller 2013). As forum sponsors, convening actors are usually more positive about forum performance (Leach & Sabatier 2005). Scientific researchers and consultants also generally attempt to adhere to a more neutral role, in which they provide information that allows decision-makers to clarify the consequences of their choices (Angst & Brandenberger 2022; Fischer & Schläpfer 2017).

In contrast, some actors take on an advocacy role in which they prefer a specific distribution of resources – for example, emphasizing green (nature-based) adaptation over grey infrastructure, or emphasizing the cost of infrastructure assets over the cost of protecting vulnerable communities. We define these actors as "partial." Chief among advocacy actors are environmental and economic interest groups (Angst & Brandenberger 2022). Local governments, water districts, and environmental districts frequently advocate for their narrower policy interests rather than play the neutral brokering role of other government actors (Ekstrom & Moser 2014) so we label them "partial" also. A total 461 respondents answered this part of the survey, though only 316 attended one of the big forums. For an outline of the different types of actors in the data used in this section, see Figure 20.

In Section 4, we argued that the leaders of the governance process in SF Bay lead the system throughout the planning cycle by providing both bonding and

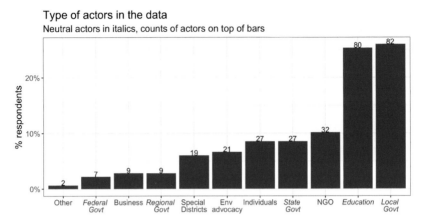

Figure 20 Types of actors in the data.

bridging social capital in the policy network and maintaining simultaneous conversations with different types of actors. Yet, this type of leadership may not be sufficient to catalyze cooperation around the political and distributional aspects that governing adaptation often entails. These pertain, first and foremost, to planning and funding. Given the high funding requirements of most adaptation projects, governmental resources are necessarily involved. Yet government does not have infinite funding capacity. Priority locations and assets need to be identified, and decisions made concerning which type of infrastructural solutions to implement. These planning decisions are rife with conflict, especially when the differential vulnerabilities and adaptive capacities of "frontline" communities require consideration of environmental and climate justice. Because regional government actors do not have authority to fully decide on these matters (because land use issues in California are decided at local level, and because the funding requirements exceed what governmental agencies could fund given their existing resources), "neutral" governmental actors and experts will stay clear of discussing them. We contend that this causes discontent in "partial" actors, particularly at advanced stages of the understanding phase of the planning cycle, because it does not address their core concerns.

H2: Partial actors will be less satisfied with the performance of the big games than neutral actors.

Yet, previous research found that actors who are well-informed about the policy issue at hand tend to perform better in polycentric systems and to reap most of the benefits from participation (Mewhirter & Berardo 2019). The role of specialized knowledge in this system has been found to lead to more positive perception of forum progress (Vantaggiato & Lubell, 2022). Thus, we expect that even partial actors will display more positive perceptions of the forums they attend if they are highly knowledgeable about SLR.

H3: The higher the actors' specialization about SLR, the more positive their assessment of the forums.

Dependent Variables

Our dependent variables are actors' perceptions of forum performance. We asked survey respondents to express their agreement with statements concerning five elements of forums' process and outcomes. For process, we asked our respondents about forum inclusivity (whether the goals of all participants are taken into account in the collaborative process) and about their assessment of the fairness of the collaborative process; for outcomes, we asked our respondents, for each forum they attended, how it affected their goals, whether it resulted in tangible progress on SLR, and whether it produced innovative thinking on how to address SLR in SF Bay. We measured inclusivity, tangible progress, and innovative thinking on a five-point Likert scale from "strongly disagree" to "strongly agree." We measured impact and fairness using a scale from 0 to 10, where 0 meant "major negative impact/very unfair" and 10 meant "major positive impact/very fair." To make these measures of forum evaluation comparable, we normalized them to fall in the range between 0 and 1 using min–max normalization (Suarez-Alvarez et al. 2012).

Explanatory Variables

The first key explanatory variable we use is a binary variable that we call *Stance* which takes value 1 if the actor is "partial" and 0 if the actor is "neutral."

The second explanatory variable measured at individual level is whether the respondent is a *Specialist of SLR*, which refers to the level of an actor's involvement in the governance system. Actors who are more deeply involved with the governance issue at hand are more likely to perceive the initiative they take part in as being successful. To create this variable, first, we divided respondents into specialists and nonspecialists, based on their answers to a survey question asking whether they deal with sea level rise as a major part of their job or only occasionally. We define those who have sea level rise as a major part of their job as "specialists" and assign them a value of 1; we define those for whom sea level rise is only part of their job as "nonspecialists" and assign them a value of 0 (zero). The survey also asked respondents to indicate how long they have dealt with sea level rise as a major part of their job, with three options: up to 1 year, between 1 and 5 years, over 5 years. We categorized these answers as ordinal scores ranging from 1 to 3, with 1 indicating up to 1 year, 2 indicating between 1 and 5 years, and 3 indicating over 5 years' experience dealing with sea level rise as a major part of one's job. Thus, we created individual scores ranging from 0 (the respondent is not a specialist of SLR) to 3 (the respondent is a specialist of SLR with over 5 years' experience).

Control Variables

The first control variable measured at individual level is a binary variable indicating whether a given actor belongs to the organization that is a *Forum convenor*, for the reason that this may lead an individual actor to provide a more positive assessment of the forum (Meier & O'Toole 2013). As shown in the previous section, the "leadership core" of the policy network of SLR in the Bay Area is composed of all "neutral" actors, that is, governmental agencies from different levels of governance and scientific experts. One of the main functions that leaders perform in this system is "convening" stakeholders. Hence, most forum convenors are governmental agencies.

The second control variable measured at individual level is the *Number of Forums attended* (maximum of five by survey design) on the basis that actors who attend more forums have more knowledge and therefore achieve more impact from the forums they attend (Mewhirter et al. 2018).

The third control variable is a measure of actors' self-assessment of how informed they feel about sea level rise, to check for the robustness of the *Specialist of SLR* variable in additional models, reported in the Appendix.

Methods

The unit of analysis is the actor–forum dyad. Therefore, actors are represented in multiple observations within the data set, dependent on the number of forums in which they participate (an actor who participates in one forum has one observation; an actor who participates in three forums has three observations; and so on). Similarly, forums are represented in multiple observations, contingent on the number of respondents that participated in that forum. This presents some analytical challenges. On the one hand, respondents may report similar experiences across forums that depend on their stance and goals concerning the governance of adaptation to sea level rise. On the other, each forum may affect respondents in similar ways because of its characteristics. To deal with interdependencies between actors' responses, we rely on multilevel regression models with crossed random effects for respondents and forums estimated using the *lmer* package in R.

Descriptive Analysis: Actors' Perceptions of the Big Games

Figure 21 summarizes our respondents' assessment of the seven big games according to our five dependent variables of forum performance: inclusivity, tangible progress, innovative thinking, fairness, and impact. The figure also indicates the main goal or purpose of each forum, which we derive from forums' stated objectives and from our qualitative familiarity with each, and comprise networking (BayCAN and CHARG), carrying out vulnerability assessments (ART, BayWAVE, and SeaChange) and planning (Resilient by Design and SR37). As shown in Figure 21, forums focused on vulnerability assessment are the best performers in all categories except "innovative thinking," where they come second after Resilient by Design. In contrast, planning forums perform worst in terms of fairness and inclusivity while networking forums perform worst in terms of impact and tangible progress.

The two planning forums score similarly on all aspects, with the exception made for "innovative thinking," where Resilient by Design scores highest of all. We attribute this difference to the very different setups of the two forums. RBD was a planning design competition which attracted leading engineering and design firms from all over the country to propose their ideas; its very purpose was providing innovative thinking. In contrast, SR37 is a subregional policy committee comprising governmental actors and local interest groups, that has been debating a handful of policy solutions since 2015, with limited progress or breakthroughs.

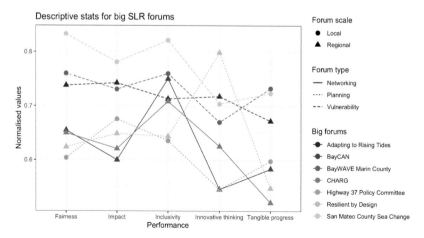

Figure 21 Descriptive statistics for the "big games.".

The findings outlined in Figure 21 lend support to our hypothesis 1, which foresees that forums focused on understanding tasks (i.e., vulnerability assessments and networking) will fare better than forums focused on planning tasks. This is because the tasks of understanding come with barriers, which a functioning polycentric governance system with high administrative capacity, like SF Bay, is well suited to overcome. In contrast, planning tasks involved a suite of barriers involving prioritization of different infrastructural projects and, therefore, locations, which have distributional implications and which a polycentric governance system may lack the political legitimacy and therefore the authority to address.

Regression Model Results

The results of the main regression models are presented in Figure 22. When studying the figure, please bear in mind that the dependent variables have all been standardized to fall within the 0 to 1 range. Tables with results of these and additional models are reported in the Appendix.

The results show several things. First, we find our second hypothesis confirmed by the fact that actors whom we defined as "partial" (i.e., partial to specific adaptation solutions) assess the forums they attend less positively than "neutral" actors. This is true for all dependent variables, except fairness. In contrast, we find that actors who are "specialists of SLR" perceive more positive impact of the forums they attend on their own organizational goals. This lends support to our contention that "neutral" actors (governmental agencies and researchers and scientists) structure debate around issues of technical knowledge and information dissemination.

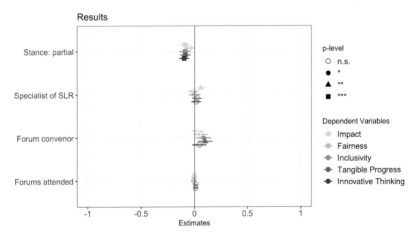

Figure 22 Actors' perceptions of forum performance.

Our control variables display both expected and unexpected results. Among the expected results, we find that forum convenors (typically, state or county agencies) are significantly more likely to assess the forums they convene as more inclusive and leading to more tangible progress than other respondents (confirming the findings in Hamilton (2018)). Unexpectedly, the number of forums attended by individual participants has no bearing on their assessment of any forum, differently from what Mewhirter et al. (2019) find – although it must be borne in mind that we capped the maximum number of forums a survey respondent could mention to five.

These results are robust to the inclusion of an alternative measurement of information about the short-term consequences of SLR (as perceived by actors themselves) as reported in the Appendix (Table A4). Though very few, actors who feel not informed about SLR are considerably more likely to evaluate the outputs of the forums they attend less positively than actors who feel very well or well informed. Results for models including actors' self-assessment of how informed they are about the long-term consequences of sea level rise paint an identical picture. In the same models, we included actors' self-reported concern for the short-term consequences of sea level rise. The variable has no effect, suggesting that information is indeed what matters for actors' perceptions rather than their general interest about the issue. In the Appendix, we also report the results of models including a breakdown of respondents by actor type (Table A3); those results show that the actors we term "partial" (environmental advocates, special districts, NGOs, individuals) are indeed those more likely to report lower evaluation of the performance of the forums they attend.

Finally, and importantly, we find no support for our third hypothesis that actors who are specialists of sea level rise would rate forums higher even when they belong to the "partial" category (results not reported, see Appendix Table A2). The interaction has no effect, while all other coefficients maintain their sign and significance. This suggests that the actors who benefited the most from forum participation are those with high levels of knowledge and no preexisting preference for the redistribution of resources within the Bay.

Discussion and Conclusions

As we mentioned in Section 1, SF Bay is culturally predisposed to collaborative governance and consensus-seeking (Calanni et al. 2015). "Getting everybody in the room" – the main challenge of policy forums – is not a challenge in the Bay Area. This feature allows us focus on the "real" challenge of climate adaptation governance: Once all interested parties are in the room, finding consensus between them proves quite hard, because although they all agree on the exist-ence of the problem and the need to address it (see the figures in Section 2), they disagree on how to do so. The findings from our interviews clarified that one of the outcomes of the understanding phase of the adaptation planning cycle is that it becomes clear to policy actors that there is a mismatch between the costs of adaptation for the whole Bay and available local, regional, and even state and federal resources. This mismatch renders adaptation a political process of resource allocation. Therefore, in addition to discussing different planning options, policy actors try to make sure their locations of interest become prioritized as recipients of funds for adaptation planning. This may well be the main rationale of most "partial" policy actors who take part in policy forums.

Yet, "partial" actors tend to report less satisfaction with their ability to meet their goals within the forums they attend; they also report less satisfaction with the progress made by forums and the innovative thinking it produced. They also are less in agreement with "neutral" actors that forums take all stakeholders' goals into account and that the decision-making process is fair.

These findings are important: The relative dissatisfaction of "partial" actors with forum performance on both outcomes and process suggests that there is also a mismatch between the capabilities of the polycentric system and the demands of the policy actors therein. This may have two potential conse-quences: the most resourceful, well-connected policy actors will bring their stances to more sympathetic forums, while others will disengage. The first outcome is undesirable as it may open avenues for political influence that are unavailable to some. The second outcome is particularly undesirable when the disengaged are disadvantaged/community-based organizations, who are likely to bear the brunt of the consequences of a changing climate.

As the most recent and central forum to emerge from the set of "big games" analyzed here, BayAdapt (launched in 2020) recognizes but remains hampered by these issues. BayAdapt was sponsored by the BCDC and developed an innovative multilevel planning process that featured coordin-ation among organizational leaders, community engagement, and broader public outreach. BCDC planners moved among these levels to integrate knowledge and preferences. Environmental justice and the inclusion of frontline communities was a central goal from the start, although some community-based groups and leaders felt they should have been included earlier and in more prominent positions both in the process and resulting documents. In other words, the policy leaders in BayAdapt had to learn over time within the forum about how to address the needs of different advocacy groups. Yet despite all of this learning, cooperation, and leadership,

BayAdapt produced a "joint platform" that did not identify specific adaptation projects or compel other actors to contribute to implementation. Thus, BayAdapt has now entered an "implementation" phase where the questions of resource distribution and cooperation become even more poignant as real money is allocated to real projects, and policy actors are asked to contribute even higher levels of their own resources. Whether BayAdapt will succeed in implementation – in conjunction with all the other existing forums and any new ones that will emerge – remains to be seen.

The story of BayAdapt exemplifies how when policy forums attempt to move the system beyond the understanding phase, resource and distributional conflicts become major barriers to adaptation. Even when relatively neutral actors such as government agencies, scientific, and knowledge brokers seek to identify and implement mutually beneficial solutions at the regional level, the distributional conflicts over resource adaptation become more severe as the adaptation planning cycles progresses. The costs and benefits of adaptation are not equally distributed across actor types or geographies, and thus naturally more "partial" actors will have divergent policy preferences and tend to be less satisfied with current collaborative solutions. This observation identifies distributional politics as a main source of barriers to adaptation, and places even more importance on collaborative governance and other conflict resolution strategies.

6 Implications and Agenda for Future Research

How do polycentric governance systems respond to a new collective action problem like climate adaptation? Our analysis of sea level rise adaptation in SF Bay paints a portrait of a governance system that is adding a new layer of institutions and networks onto an already complex ecology of environmental governance games that developed over many decades. Environmental policy actors focused on issues like ecosystem and infrastructure management are now grappling with how sea level rise will impact their core policy goals. In response, they form new networks, build new policy forums, or shift existing policy forums to enable learning, cooperation, and bargaining over possible sea level rise adaptation scenarios. SF Bay remains in the midst of this process of institutional change.

However, such institutional change is "easier said than done" because addressing the emerging collective action problem of climate change requires overcoming the barriers and challenges identified in research on the adaptation planning cycle. Our evidence suggests that the SF Bay system remains mostly in the understanding and planning stage of the adaptation planning cycle, with only minimal progress to the implementation and management stage, which would feature widespread implementation and monitoring of on-the-ground adaptation projects. Our argument is that as the system moves toward later stages of the adaptation planning cycle, then the political questions of bargaining over resource distribution become much more difficult to address. Concomitantly, while environmental justice and equity questions are present from the start, they become even more poignant at later stages of the process when some communities are disadvantaged in improving their adaptive capacity (Chu & Cannon 2021; Holland 2017; Paavola 2008). In other words, the transaction costs of collective action increase as the system progresses through the adaptation planning cycle. As a result, adaptation is typically incremental and policy actors are justified in worrying about whether adaptation can keep pace with the rate of environmental change, even for "slow moving emergencies" like sea level rise.

These processes of institutional change also illustrate what we might call the "democratic dilemma" of polycentric systems: Everybody wants to be involved in developing an overarching regional plan for climate change adaptation, but nobody wants to concede political power or confer additional authority to an existing or new agency to lead the development and implementation of the plan. As California water policy veteran Phil Isenberg (Isenberg 2016) commented regarding polycentric systems, "Public policy is almost always a mess. Let's acknowledge the inevitable and figure out how to manage a messy situation. Trying to define a policy 'problem' is hard enough. Trying to find a solution is even harder. Trying to do either in a policy-making structure in which everyone is involved, but nobody is in charge, is nearly impossible." The extent to which polycentric systems can successfully navigate the horns of this democratic dilemma is key to the effectiveness of climate adaptation and has implications for four bodies of literature: adaptation planning cycles, polycentric governance, collaborative governance, and broader theories of the policy process. The first two of these literatures form the main theoretical basis for this Elements, but we think our findings are also germane to collaborative governance and broader theories of the policy process that could also be applied to the analysis of climate adaptation.

Implications for the Adaptation Planning Cycle Literature

Barriers to adaptation are a core concern of the adaptation planning cycle literature. Barriers that originate from different sources apply to different stages of the planning cycle, and collectively work to stall progress. Depending on the region or organizational unit being studied, the literature finds that some organizations or regions are further along the adaptation planning cycle than others. The practical question then becomes diagnosing the sources of the barriers and designing effective strategies to overcome them. Adaptation planning can fail when the pace of the climate change impacts is faster than the incremental progress in the climate adaptation planning cycle, or when truly transformational adaptation is needed rather than incremental progress (Barnes et al. 2017; Herrfahrdt-Pähle et al. 2020).

In this study, we framed the adaptation planning cycle within the processes theorized in the EGF, with its emphasis on learning, cooperation, and bargaining as the social processes fueling system evolution. Our study illuminates how the adaptation planning cycle is integrated with the process of institutional change in polycentric systems. In the early *understanding* phase, processes of learning make actors feel empowered, provide opportunities for institutional creation, for broad engagement of different societal actors, and for the dissemination of high-quality information on the problem. At the time of our observation, the polycentric system had reached a clear understanding of the vulnerabilities (learning) and was focused on promoting a regional approach with stakeholders (cooperation). The *planning* phase involves making decisions with distributional implications concerning how and where to spend scarce resources (bargaining); in essence, whom to protect first and how much to protect them. We observed this social process in the cases of planning policy forums such as RBD and SR 37 (see Section 5), which attracted a lot of controversy and conflict. It is in the planning stage that the political conflicts over the distribution of the costs and benefits of adaptation, as well as the potential procedural and distributional equity issues, become more serious barriers to progress. We think this occurs because while equity and bargaining are present from the start, the material consequences of bargaining and resource distribution are more directly experienced as the adaptation planning cycle advances, which incentivizes policy actors to fight harder for their policy preferences.

Overall, climate adaptation in the context of polycentric systems highlights the important role of politics throughout the climate adaptation planning cycle. Our reading of the literature on adaptation planning is that it devotes little attention to the "political" aspects of governance processes. Yet, politics matters, and political conflict becomes more severe as the adaptation planning cycle moves forward. This increases overall transaction costs, rendering the barriers more difficult to overcome. The fact that consensus-building and collaboration are ultimately saddled and slowed down by conflict between diverging interests even in a policy context with ample familiarity with polycentric and collaborative governance, such as the Bay Area, suggests that climate adaptation is a political issue, with a big P (Javeline 2014). Indeed, recent scholarship has highlighted the political bottlenecks of environmental problems and emphasized the distributional implications of climate adaptation decisions (Ciplet et al. 2013; Dolšak & Prakash 2018) as well as the poignancy of environmental justice (Chu & Cannon 2021; Holland 2017; Khan et al. 2020) to any conversation concerning planning for adaptation.

Implications for the Polycentric Governance Literature

The literature on polycentric governance is divided into two "camps": enthusiasts and skeptics. Enthusiasts highlight the benefits of polycentric systems, which are portrayed as more inclusive, more informed, and more flexible than "monocentric" ones. Policy actors benefit from the many centers of deliberation and decision-making that characterize polycentric systems, which offer multiple opportunities for engagement, learning, and pursuing policy goals (Mewhirter & Berardo 2019). The enthusiasts mostly dominated the earlier literature on polycentric governance, to such an extent that polycentric governance has become the normative "prescription" for good governance that can promote multilevel cooperation and adapt to uncertainty over time.

However, more skeptical perspectives have emerged to challenge this rosy picture and emphasize the shortcomings of polycentric governance. Skeptics underline that whatever the characteristics of polycentric systems, these do not affect the preexisting characteristics of governance actors, and hence the power dynamics that exist outside of polycentric systems are bound to be reproduced within them (Eriksen et al. 2015; Morrison et al. 2019). As for inclusion, the openness of polycentric systems does not imply that all affected actors are able to take part in deliberative processes: Actors have different resources and capabilities to participate in the multiple conversations that occur in the polycentric system, which often do not go far enough to ensure that those who are "disadvantaged" get an actual voice in the policy process (Dobbin & Lubell 2019; Dobbin et al. 2023). Moreover, the relative informality of polycentric systems often entails the absence of provisions for transparency and accountability for any decisions made (Bäckstrand et al. 2018). The characteristics of polycentric systems, with multiple actors, issues, and forums, contribute to fragmentation and increase the transaction costs of cooperation.

Our study speaks to both sides of this debate. On the one hand, we recognize the structural signs of leadership combined with more localized cooperative relationships between different actors (Section 4) which suggest the polycentric governance system achieves coordination while featuring multiple centers of local interaction; on the other hand, we find evidence that the governance system is skewed toward those who possess the expertise and knowledge to engage productively, but it does not squarely address the concerns of those who advocate for different solutions (Section 5). Thus, we see both evidence of the purported benefits of polycentric governance and of its shortcomings. We propose that we acknowledge that climate governance is polycentric and start asking what polycentricity is good for, that is, what collective action problems can it solve? This is the thrust of the EGF and its focus on studying which types of forums and relationships can disentangle the collective action "knots" in the governance process.

Our study shows that polycentric governance systems can facilitate coordination by convening a wide variety of actors, informing them, and framing the issue of sea level rise as a regional problem requiring compromise and coordination. Our study also suggests that polycentric systems struggle to convincingly address questions of redistribution and justice. We surmise that this is because neither the leaders nor the actors in the core of the network have the authority to address those questions. In the case of SF Bay, polycentricity, however, allowed issues of redistribution and justice to be mainstreamed in the governance system. Addressing these issues requires political legitimacy and authority which, at the time of our observation, no organization or actor could fully claim. The BayAdapt planning process is probably the key feature of the system to observe going forward, because it features a lead agency (BCDC) trying to develop a regional vision and implementation plan. But even at the time of this writing, new policy games are still emerging, and the system is quickly changing.

Implications for the Collaborative Governance Literature

While the EGF and polycentric governance take a systems-approach that considers multiple policy games, actors, and issues, the literature on collaborative governance focuses on the key principles and variables that influence the evolution of cooperation within a particular policy forum. For example, Emerson et al. (2015) posit that collaboration is a function of principled engagement, capacity for joint action, and shared motivation.

Ansell and Gash (2008) focus on how initial starting conditions such as power asymmetries and history of cooperation influence the process of collaboration including trust-building, commitment, face-to-face dialog, intermediate outcomes, and building a shared understanding.

Our study shows that adaptation barriers are related to many of the variables identified by the collaborative governance literature. For example, the lack of a regional plan, resource deficiencies, permitting obstacles, uncertainty, and institutional fragmentation are starting conditions that constrain collaboration across the system. In terms of process, respondents pointed to lack of a common vision, distrust, leadership, and procedural injustice as barriers throughout the system. However, because climate adaptation features polycentric systems, it is important to think about the extent to which these constraints play out differently over time or across forums, or whether every forum must achieve the same collaborative governance principles. For example, there are some forums that specialize more in science to reduce uncertainty while others focus more on community engagement and environmental justice. Hence, if these various forums are integrated by networks of policy actors, the system as a whole may approach the principles of collaborative governance even if not every single forum is collaborative.

Furthermore, while many policy forums reflect collaborative design principles, not all policy forums are collaborative – regulatory processes and forums with closed membership also exist in the system. Many of these forums involve conflict; indeed more conflict occurs in the "big games" (Lubell et al. 2020), low-cost conflict resolution forums are an important contributor to cooperation (Ostrom 2010b), and there may be positive feedbacks between conflict and cooperation (Weible & Heikkila 2017). Hence, polycentric governance systems are truly a case of institutional diversity (Ostrom 2010a) and how such diversity changes over time is a key research question.

Learning is a key feature of theories of collaborative governance (Heikkila & Gerlak 2019; Pahl-Wostl 2009), as well as the early stages of the adaptation planning cycle (Teodoro et al. 2021) and the evolution of new policy games in polycentric systems (Berardo & Lubell 2016; Vantaggiato & Lubell 2020). This literature emphasizes the positive impact of learning on participants' perceptions of the quality of the process; as participants are exposed to accurate and accessible information on the practical implications of the environmental policy problem/issue, they change their beliefs and achieve consensus (Calanni et al. 2015). However, this study shows that learning does not necessarily lead to action. While actors in SF Bay report having learned from forum participation, we also find that learning matters most for those who are specialists of SLR already and for those that we called "neutral", that is, actors who would like to achieve coordination and a regional approach but have no predetermined agenda (i.e., governmental agency officers and experts of various kind). Advocacy actors are less positive than neutral actors on the performance of the system. More generally, we see learning occurring and being positively received by actors, but little on-the-ground implementation.

Implications for the Theories of the Policy Process

Our analysis has integrated two theoretical approaches – the ecology of games applied to polycentric governance and the adaptation planning cycle – to analyze sea level rise adaptation in SF Bay. However, other theories of the policy process can also serve as a useful lens for analysis of these governance processes. Our results have implications for two other theories of the policy process in particular: the advocacy coalition framework (ACF) and punctuated equilibrium theory.

The ACF emphasizes conflict between coalitions that are held together by shared belief systems. Sea level rise in SF Bay is an example of a nascent subsystem (Gmoser-Daskalakis et al. 2023). The SF Bay Area has the advantage of a political culture that is generally supportive of climate change policies. Hence, there is not a strong pro-SLR versus anti-SLR coalition, where the latter denies the existence of SLR or the necessity of finding adaptation pathways. However, there are emergent coalitional dynamics around environmental versus community and environmental justice values. Environmental groups seek to ensure that any local adaptation solutions have environmental cobenefits or at least avoid environmental harms. This is one reason they support green infrastructure solutions. These environmental values may come into conflict with communities that are seeking to protect critical infrastructure and community assets, regardless of whether any adaptation strategies may compromise environmental values. These coalitions may become more crystallized over time as the subsystem matures and moves into later phases of the adaptation planning cycle, where distributional conflicts become more salient. With its focus on emergence and coordination, the EGF provides a suitable theoretical and empirical background for further study of "nascency" in policy subsystems.

Punctuated equilibrium theory analyzes how multiple streams of the policy process may converge to elevate a policy issue like SLR from a parallel process at lower levels of a complex governance system to serial processing at more central policy forums. SLR has become a more salient issue over time, although arguably the discussion around SLR began in regional forums that were addressing other environmental issues like water management (Lubell & Robbins 2021). But as awareness of SLR continued to increase, many local government units and other types of public and private organizations began to act in a mostly uncoordinated and fragmented manner. This has created a demand for more leadership at the regional level, as witnessed by the BCDC's convening of the BayAdapt regional planning process. However, implementation of BayAdapt remains a considerable challenge and it is probably fair to say that the regional governance system remains in an incremental rather than punctuated or transformational mode of institutional change.

Future Research

The research presented here is inherently limited because it focuses on a single case study region and the quantitative data only captures a single snapshot in time. The qualitative research, in particular our ongoing policy engagement, enables us to observe some change over time but even that is limited. To really understand how polycentric systems respond to emerging collective action problems, we need much more research over time, space, and for different types of issues. In addition, we need to develop a deeper understanding of the role of political power and equity in the context of polycentric governance.

The evolution of polycentric governance systems occurs over long periods of time and is path dependent (Bell & Olivier 2021; Weible et al. 2020). But we do not know whether the systems evolve in linear or nonlinear fashion, and how different social processes cycle over time. From the SF Bay context, we know that climate adaptation institutions emerged from a rich ecology of environmental governance games that grew in the Bay Area over decades. For example, many of the actors involved in SLR were also involved in the collaboration networks associated with integrated regional water management (Lubell & Robbins 2021). Sea level rise and climate adaptation experienced a rapid increase in salience and associated growth in institutions that resembles a punctuated equilibrium. But we are not sure whether this growth will continue, or level off, or decline, for example, as with issue-attention cycles (Downs 1991) or the lifecycle seen in collaborative governance (Ulibarri et al. 2023).

At the microlevel, we also do not know how the institutions will evolve with cycles of cooperation, conflict, learning, and bargaining.

Different social processes may be more prominent at different points in time, or in different regions of the governance "brain" formed by polycentric institutions. Hence, just like long-term ecological research stations, we need long-term social–ecological research infrastructure to track changes over decades.

Furthermore, the evolution of polycentric governance systems will vary across space in different regions and countries. There are many candidates for contextual variables that will shape the evolutionary process: political culture (e.g., extent of climate denial), macro-political institutions (e.g., authoritarian or democratic), nature of the problem (e.g., severity of flooding), and resources (e.g., budgets) among others. The constellation of contextual variables will influence how quickly polycentric institutions will recognize and respond to new collective action problems, the permanence of any institutional changes, and their capacity to deliver policy outputs and outcomes that address the problems. Some polycentric systems never even reach the problem identification stage, let alone go through the full adaptation planning cycle and even respond to feedbacks. Indeed, our research suggests that even in a region like SF Bay, which has many advantages with respect to climate adaptation governance, it is easy to get stuck.

While our research makes the point that climate adaptation involves collective action problems, there are obviously many more collective action problems being addressed by polycentric institutions. Climate mitigation and other environmental collective action problems are well known, and certainly have witnessed the evolution of polycentric governance. But in our view, basically all policy issues involve interdependencies across actors and thus collective action problems. We think our empirical and theoretical approach would be especially useful in the context of public health, crime, and education research, although really any policy issue where policy networks and cooperation matter are good candidates. As we have mentioned here and in other papers (Morrison et al. 2023) – everything is polycentric! We need to understand how the structure and function of polycentric governance operates for different issues, across time and space.

Lastly, we need a better understanding of political power and equity in the context of polycentric governance and ecology of games. Morrison et al. (2019) have made some important contributions to these questions, but more empirical work is needed. The EGF suffers from its roots in transaction cost economics, which is more concerned about the efficiency of cooperation rather than procedural and distributional justice. However, the framework does include the concept of bargaining as an opportunity to understand how policy actors may exert political power to pursue their desired outcomes. But more specific analysis is needed regarding procedural (e.g., representation and political efficacy in policy forums) and distributional justice (e.g., which groups receive more benefits of climate adaptation), and how different actors such as environmental justice organizations may self-organize as advocacy coalitions.

Appendix

Bayesian goodness-of-fit diagnostics

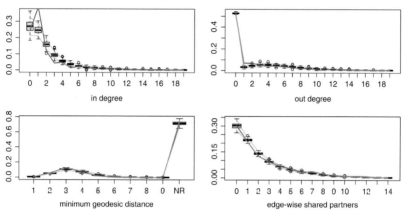

Figure 1A Goodness of fit diagnostics from BERGM in Section 4.

Table A1 Results of models shown in Section 5.

	Impact	Fairness	Inclusivity	Tangible progress	Innovation
Intercept	0.69 (0.03)***	0.71 (0.04)***	0.75 (0.04)***	0.59 (0.04)***	0.64 (0.05)***
Partial	−0.09 (0.02)***	−0.05 (0.03)˙	−0.09 (0.03)***	−0.09 (0.02)***	−0.10 (0.02)***
Specialist of SLR	0.06 (0.02)*	−0.01 (0.03)	0.00 (0.03)	0.02 (0.02)	0.02 (0.02)
Forum convenor	0.03 (0.03)	0.06 (0.03)˙	0.09 (0.04)*	0.10 (0.03)**	0.05 (0.03)
Forums attended	−0.00 (0.01)	−0.00 (0.01)	−0.01 (0.01)	0.01 (0.01)	0.01 (0.01)
AIC	−175.90	−97.31	7.37	−67.18	−85.63
BIC	−142.90	−64.54	40.42	−34.63	−52.72
Log likelihood	95.95	56.65	4.31	41.59	50.82
Num. obs.	457	444	460	432	452
Num. groups: ResponseID	290	287	297	288	299
Num. groups: forum	7	7	7	7	7
Var: resp (intercept)	0.02	0.02	0.02	0.01	0.01
Var: forum (intercept)	0.00	0.00	0.00	0.01	0.01
Var: residual	0.02	0.03	0.04	0.03	0.03

*** $p < 0.001$; ** $p < 0.01$; * $p < 0.05$; ˙ $p < 0.1$

Table A2 Models testing H2 of Section 5 (interaction)

	Impact	Fairness	Inclusivity	Tangible progress	Innovation
Intercept	0.69 (0.04)***	0.72 (0.04)***	0.75 (0.04)***	0.59 (0.04)***	0.64 (0.05)***
Partial	−0.10 (0.03)**	−0.06 (0.03)·	−0.09 (0.04)*	−0.08 (0.03)*	−0.08 (0.03)*
Specialist of SLR	0.05 (0.03)*	−0.02 (0.03)	−0.00 (0.03)	0.03 (0.03)	0.03 (0.03)
Forum convenor	0.03 (0.03)	0.06 (0.03)·	0.09 (0.04)*	0.10 (0.03)**	0.04 (0.03)
Forums attended	−0.00 (0.01)	−0.00 (0.01)	−0.01 (0.01)	0.01 (0.01)	0.01 (0.01)
Stance * specialization	0.01 (0.05)	0.02 (0.05)	0.00 (0.05)	−0.02 (0.05)	−0.04 (0.05)
AIC	−169.66	−91.33	13.34	−61.18	−80.21
BIC	−132.54	−54.47	50.52	−24.57	−43.19
Log likelihood	93.83	54.66	2.33	39.59	49.11
Num. obs.	457	444	460	432	452
Num. groups: ResponseID	290	287	297	288	299
Num. groups: forum	7	7	7	7	7
Var: resp (intercept)	0.02	0.02	0.02	0.01	0.01
Var: forum (intercept)	0.00	0.00	0.00	0.01	0.01
Var: residual	0.02	0.03	0.04	0.03	0.03

*** $p < 0.001$; ** $p < 0.01$; * $p < 0.05$; · $p < 0.1$

Table A3 Models with breakdown by type of actor (Section 5).

	Impact	Fairness	Inclusivity	Tangible progress	Innovation
Intercept	0.69 (0.04)***	0.70 (0.05)***	0.73 (0.05)***	0.58 (0.05)***	0.66 (0.06)***
Regional govt (federal, state, regional)	Reference category				
Business	0.03 (0.08)	0.09 (0.08)	0.08 (0.09)	0.08 (0.08)	0.03 (0.08)
Education	0.05 (0.04)	0.02 (0.04)	0.04 (0.04)	0.03 (0.04)	0.01 (0.04)
Env advocacy	−0.18 (0.05)***	−0.02 (0.06)	−0.07 (0.06)	−0.14 (0.05)*	−0.12 (0.05)*
Local govt	−0.06 (0.03)	−0.01 (0.04)	−0.01 (0.04)	−0.03 (0.04)	−0.04 (0.04)
NGO	−0.09 (0.05)*	−0.07 (0.05)	−0.12 (0.05)*	−0.06 (0.05)	−0.10 (0.05)*
Other	−0.07 (0.12)	−0.16 (0.14)	−0.33 (0.15)*	−0.25 (0.13)	−0.16 (0.13)
Individuals	−0.10 (0.05)*	−0.01 (0.05)	−0.02 (0.06)	−0.09 (0.05)	−0.15 (0.05)**
Special districts	−0.07 (0.05)	−0.11 (0.06)	−0.11 (0.06)	−0.09 (0.05)	−0.12 (0.05)*
Specialist of SLR	0.04 (0.02)	−0.03 (0.03)	−0.01 (0.03)	0.02 (0.02)	0.01 (0.02)
Forum convenor	0.04 (0.03)	0.07 (0.03)*	0.09 (0.04)*	0.11 (0.03)**	0.05 (0.03)
Forums attended	0.00 (0.01)	0.00 (0.01)	−0.00 (0.01)	0.02 (0.01)	0.01 (0.01)
AIC	−153.47	−63.98	36.66	−36.72	−50.75
BIC	−91.60	−2.54	98.63	24.31	10.96
Log likelihood	91.74	46.99	−3.33	33.36	40.37

Num. obs.	457	444	460	432	452
Num. groups: ResponseID	290	287	297	288	299
Num. groups: forum	7	7	7	7	7
Var: resp (intercept)	0.01	0.02	0.02	0.01	0.01
Var: forum (intercept)	0.00	0.00	0.00	0.01	0.01
Var: residual	0.02	0.03	0.04	0.03	0.03

$^{***}p < 0.001$; $^{**}p < 0.01$; $^{*}p < 0.05$; $^{\cdot}p < 0.1$

Table A4 Results with perceptions of short-term information and concern (Section 5).

	Impact	Fairness	Inclusivity	Tangible progress	Innovation
Intercept	0.67 (0.05)***	0.69 (0.05)***	0.74 (0.05)***	0.60 (0.05)***	0.63 (0.06)***
Stance: partial	−0.09 (0.02)***	−0.05 (0.03)	−0.08 (0.03)**	−0.08 (0.02)**	−0.09 (0.02)***
Specialist of SLR	0.06 (0.02)*	−0.01 (0.03)	−0.01 (0.03)	0.01 (0.03)	0.01 (0.02)
Forum convenor	0.03 (0.03)	0.07 (0.03)	0.09 (0.04)*	0.11 (0.04)**	0.05 (0.03)
Forums attended	0.00 (0.01)	−0.00 (0.01)	−0.01 (0.01)	0.01 (0.01)	0.01 (0.01)
Very well-informed	Reference category				
Well-informed	0.02 (0.03)	0.01 (0.03)	−0.01 (0.03)	−0.00 (0.03)	−0.01 (0.03)
Somewhat informed	0.03 (0.03)	0.03 (0.04)	−0.01 (0.04)	−0.02 (0.03)	−0.03 (0.03)
Not informed	−0.02 (0.11)	−0.01 (0.12)	−0.25 (0.11)*	−0.32 (0.10)**	−0.28 (0.12)*
Very concerned	Reference category				
Concerned	0.00 (0.03)	0.01 (0.03)	0.04 (0.03)	0.03 (0.03)	0.05 (0.03)*
Somewhat concerned	−0.04 (0.03)	−0.00 (0.03)	0.01 (0.03)	−0.02 (0.03)	0.03 (0.03)
Not concerned	−0.06 (0.06)	−0.03 (0.07)	−0.07 (0.07)	−0.08 (0.07)	0.00 (0.06)
AIC	−137.75	−57.78	37.42	−42.89	−54.64
BIC	−80.09	−0.53	95.16	13.97	2.86
Log likelihood	82.87	42.89	−4.71	35.44	41.32
Num. obs.	454	441	457	429	449

Num. groups: ResponseID	288	285	295	286	297
Num. groups: forum	7	7	7	7	7
Var: resp (intercept)	0.02	0.02	0.02	0.01	0.01
Var: forum (intercept)	0.00	0.00	0.00	0.01	0.01
Var: residual	0.02	0.03	0.04	0.03	0.03

$^{***}p < 0.001$; $^{**}p < 0.01$; $^{*}p < 0.05$; $^{.}p < 0.1$

References

Adger, W. Neil. 2003. "Social Capital, Collective Action, and Adaptation to Climate Change." *Economic Geography* 79(4): 387–404.

Ahlquist, John S., and Margaret Levi. 2011. "Leadership: What It Means, What It Does, and What We Want to Know about It." *Annual Review of Political Science* 14: 1–24.

Ajzen, Icek et al. 2011. "Knowledge and the Prediction of Behavior: The Role of Information Accuracy in the Theory of Planned Behavior." *Basic and Applied Social Psychology* 33(2): 101–117.

Allan, Jen I., Charles B. Roger, Thomas N. Hale, Steven Bernstein, et al. 2023. "Making the Paris Agreement: Historical Processes and the Drivers of Institutional Design." *Political Studies* 71: 914–934. https://doi.org/10.1177/00323217211049294.

Angst, Mario, and Laurence Brandenberger. 2022. "Information Exchange in Governance Networks: Who Brokers across Political Divides?" *Governance* 35(2): 585–608.

Angst, Mario, Alexander Widmer, Fischer Manuel, et al. 2018. "Connectors and Coordinators in Natural Resource Governance: Insights from Swiss Water Supply." *Ecology and Society* 23(2).

Ansell, Chris, and Allison Gash. 2008. "Collaborative Governance in Theory and Practice." *Journal of Public Administration Research and Theory* 18(4): 543–571.

Bäckstrand, Karin, Fariborz Zelli, and Philip Schleiffer. 2018. "Legitimacy and Accountability in Polycentric Climate Governance." In *Governing Climate Change*, ed. Dave Huitema, Andrew Jordan, Harro van Asselt and Johanna Forster. Cambridge, UK: Cambridge University Press, 338–356.

Barabási, Albert-László. 2009. "Scale-Free Networks: A Decade and Beyond." *Science* 325(5939): 412–413.

Barnes, Michele, Örjan Bodin, Angela M. Guerrero, et al. 2017. "The Social Structural Foundations of Adaptation and Transformation in Social–Ecological Systems." *Ecology and Society* 22(4).

Baumgartner, Frank R., and Bryan D. Jones. 1993. *Agendas and Instability in American Politics*. Chicago: University of Chicago Press.

Beagle, Julie, Jeremy Lowe, Katie McKnight, et al. 2019. "San Francisco Bay Shoreline Adaptation Atlas: Working with Nature to Plan for Sea Level Rise Using Operational Landscape Units." SFEI Contribution.

Bednar, Danny, Daniel Henstra, and Gordon McBean. 2019. "The Governance of Climate Change Adaptation: Are Networks to Blame for the Implementation Deficit?" *Journal of Environmental Policy & Planning* 21(6): 702–717.

Bell, Emily V., and Tomás Olivier. 2022. "Following the Paper Trail: Systematically Analyzing Outputs to Understand Collaborative Governance Evolution." *Journal of Public Administration Research and Theory* 32(4): 671–684. https://doi.org/10.1093/jopart/muab054.

Berardo, Ramiro, and Mark Lubell. 2016. "Understanding What Shapes a Polycentric Governance System." *Public Administration Review* 76(5): 738–751.

Berardo, Ramiro, and John T. Scholz. 2010. "Self-Organizing Policy Networks: Risk, Partner Selection, and Cooperation in Estuaries." *American Journal of Political Science* 54(3): 632–649.

Bernauer, Thomas. 2013. "Climate Change Politics." *Annual Review of Political Science* 16(1): 421–448.

Biesbroek, G. Robbert, Judith Klostermann, Catrien Termeer, et al. 2011. "Barriers to Climate Change Adaptation in the Netherlands." *Climate Law* 2(2): 181–199.

Biesbroek, G. Robbert, and Alexandra Lesnikowski. 2018. "Adaptation: The Neglected Dimension of Polycentric Climate Governance?" In *Governing Climate Change: Polycentricity in Action?*, ed. Andrew Jordan, Dave Huitema, Harro van Asslet, et al. Cambridge: Cambridge University Press, 303–319.

Biesbroek, G. Robbert, Catrien J. Termeer, et al. 2014. "Rethinking Barriers to Adaptation: Mechanism-Based Explanation of Impasses in the Governance of an Innovative Adaptation Measure." *Global Environmental Change* 26: 108–118.

Bockarjova, Marija, and Linda Steg. 2014. "Can Protection Motivation Theory Predict Pro-Environmental Behavior? Explaining the Adoption of Electric Vehicles in the Netherlands." *Global Environmental Change* 28: 276–288.

Bodin, Örjan. 2017. "Collaborative Environmental Governance: Achieving Collective Action in Social-Ecological Systems." *Science* 357 (6352). https://doi.org/10.1126/science.aan1114.

Bodin, Örjan, and Beatrice I. Crona. 2009. "The Role of Social Networks in Natural Resource Governance: What Relational Patterns Make a Difference?" *Global Environmental Change* 19(3): 366–374.

Bollens, Scott A. 1986. "Examining the Link between State Policy and the Creation of Local Special Districts." *State & Local Government Review* 18(3): 117–124.

Burt, Ronald S. 2005. *Brokerage and Closure: An Introduction to Social Capital.* Oxford: Oxford University Press.

Burt, Ronald S., Ray E. Reagans, Hagay C. Volvovsky, et al. 2021. "Network Brokerage and the Perception of Leadership." *Social Networks* 65: 33–50.

Caimo, Alberto, and Nial Friel. 2011. "Bayesian Inference for Exponential Random Graph Models." *Social Networks* 33(1): 41–55.

Calanni, John C., Saba N. Siddiki, Christopher M. Weible, et al. 2015. "Explaining Coordination in Collaborative Partnerships and Clarifying the Scope of the Belief Homophily Hypothesis." *Journal of Public Administration Research and Theory* 25(3): 901–927.

Carlisle, Keith, and Rebecca L. Gruby. 2019. "Polycentric Systems of Governance: A Theoretical Model for the Commons." *Policy Studies Journal* 47(4): 927–952. https://doi.org/10.1111/psj.12212.

Chu, Eric K., and Clare E. B. Cannon. 2021. "Equity, Inclusion, and Justice as Criteria for Decision-Making on Climate Adaptation in Cities." *Current Opinion in Environmental Sustainability* 51: 85–94.

Ciplet, David, Timmons Roberts, Mizan Khan, et al. 2013. "The Politics of International Climate Adaptation Funding: Justice and Divisions in the Greenhouse." *Global Environmental Politics* 13(1): 49–68.

Coleman, James S. 1988. "Social Capital in the Creation of Human Capital." *American Journal of Sociology* 94: S95–120.

Davis, Allison, Burleigh B. Gardner, Mary R. Gardner, et al. 1941. *Deep South.* Chicago: University of Chicago Press.

Davis, Jody L., Jeffrey D. Green, Allison Reed, et al. 2009. "Interdependence with the Environment: Commitment, Interconnectedness, and Environmental Behavior." *Journal of Environmental Psychology* 29(2): 173–180.

Desmarais, Bruce A., and Skyler J. Cranmer. 2012a. "Statistical Mechanics of Networks: Estimation and Uncertainty." *Physica A: Statistical Mechanics and its Applications* 391(4): 1865–1876.

Desmarais, Bruce A., and Skyler J. Cranmer. 2012b. "Micro-Level Interpretation of Exponential Random Graph Models with Application to Estuary Networks." *Policy Studies Journal* 40(3): 402–434.

Dobbin, Kristin B., and Mark Lubell. 2019. "Collaborative Governance and Environmental Justice: Disadvantaged Community Representation in California Sustainable Groundwater Management." *Policy Studies Journal* 49(2): 562–590.

Dobbin, Kristin B., Michael Kuo, Mark Lubell, et al. 2023. "Drivers of (in) Equity in Collaborative Environmental Governance." *Policy Studies Journal* 51(2): 375–395.

Dolšak, Nives, and Aseem Prakash. 2018. "The Politics of Climate Change Adaptation." *Annual Review of Environment and Resources* 43(1): 317–341.

Downs, Anthony. 1972. "Up and Down With Ecology: The 'Issue-Attention Cycle'". *The Public Interest*, 1972.

Eisenack, Klaus et al. 2014. "Explaining and Overcoming Barriers to Climate Change Adaptation." *Nature Climate Change* 4(10): 867–872.

Ekstrom, Julia A., and Susanne C. Moser. 2014. "Identifying and Overcoming Barriers in Urban Climate Adaptation: Case Study Findings from the San Francisco Bay Area, California, USA." *Urban Climate* 9: 54–74.

Emerson, Kirk, and Andrea K. Gerlak. 2014. "Adaptation in Collaborative Governance Regimes." *Environmental Management* 54(4): 768–781.

Eriksen, Siri H. et al. 2015. "Reframing Adaptation: The Political Nature of Climate Change Adaptation." *Global Environmental Change* 35: 523–533.

Fischer, Manuel, and Philip Leifeld. 2015. "Policy Forums: Why Do They Exist and What Are They Used For?" *Policy Sciences* 48(3): 363–382.

Fischer, Manuel, and Simon Maag. 2019. "Why Are Cross-Sectoral Forums Important to Actors? Forum Contributions to Cooperation, Learning, and Resource Distribution." *Policy Studies Journal* 47(1): 114–137.

Fischer, Manuel, and Isabelle Schläpfer. 2017. "Metagovernance and Policy Forum Outputs in Swiss Environmental Politics." *Environmental Politics* 26(5): 870–892.

Folke, Carl, Thomas Hahn, Per Olsson et al. 2005. "Adaptive Governance of Social–Ecological Systems." *Annual Review of Environment and Resources* 30: 441–473.

Folke, Carl, Stephen R. Carpenter, Brian Walker, et al. 2010. "Resilience Thinking: Integrating Resilience, Adaptability and Transformability." *Ecology and Society* 15(4).

Gmoser-Daskalakis, Kyra et al. 2023. "An Item Response Approach to Sea-Level Rise Policy Preferences in a Nascent Subsystem." *Review of Policy Research* 40(6): 972–1003.

Granovetter, Mark S. 1973. "The Strength of Weak Ties." *American Journal of Sociology* 78(6): 1360–1380.

Groen, Lisanne et al. 2022. "Re-Examining Policy Stability in Climate Adaptation through a Lock-in Perspective." *Journal of European Public Policy* 30(3): 488–512.

Hamilton, Matthew. 2018. "Understanding What Shapes Varying Perceptions of the Procedural Fairness of Transboundary Environmental Decision-Making Processes." *Ecology and Society* 23(4). https://doi.org/10.5751/ES-10625-230448.

Hamilton, Matthew, and Mark Lubell. 2017. "Collaborative Governance of Climate Change Adaptation across Spatial and Institutional Scales." *Policy Studies Journal* 46(2): 222–247.

Hamilton, Matthew, Mark Lubell, and Emilinah Namaganda. 2018. "Cross-Level Linkages in an Ecology of Climate Change Adaptation Policy Games." *Ecology and Society* 23(2). https://doi.org/10.5751/ES-10179-230236.

Handcock, M., and Krista J. Gile. 2007. Center for Statistics and the Social Sciences "Modeling Social Networks with Sampled or Missing Data."

Heikkila, Tanya, and Andrea K. Gerlak. 2019. "Working on Learning: How the Institutional Rules of Environmental Governance Matter." *Journal of Environmental Planning and Management* 62(1): 106–123.

Henry, Adam Douglas. 2012. "Survey-Based Measurement of Public Management and Policy Networks." *Journal of Policy Analysis and Management* 31(2): 432–452.

Douglas Henry, Adam, Mark Lubell, and Michael McCoy. 2011. "Belief Systems and Social Capital as Drivers of Policy Network Structure: The Case of California Regional Planning." *Journal of Public Administration Research and Theory* 21(3): 419–444.

Herrfahrdt-Pähle, Elke et al. 2020. "Sustainability Transformations: Socio-Political Shocks as Opportunities for Governance Transitions." *Global Environmental Change* 63: 102097.

Hinkel, Jochen et al. 2018. "The Ability of Societies to Adapt to Twenty-First-Century Sea-Level Rise." *Nature Climate Change* 8(7): 570–578.

Holland, Breena. 2017. "Procedural Justice in Local Climate Adaptation: Political Capabilities and Transformational Change." *Environmental Politics* 26(3): 391–412.

Huitema, Dave et al. 2016. "The Governance of Adaptation Choices, Reasons, and Effects. Introduction to the Special Feature." *Ecology and Society* 21(3).

Hummel, Michelle A. 2018a. "Sea Level Rise Impacts on Wastewater Treatment Systems Along the U.S. Coasts." *Earth's Future* 6(4): 622–633.

Hummel, Michelle et al. 2018b. "Regional Adaptation to Sea Level Rise in the San Francisco Bay Area: Establishing Interdependence and Motivating Coordinated Action." paper presented at the American Geophysical Union, Fall Meeting 2018.

Hummel, Michelle A. et al. 2018c. "Clusters of Community Exposure to Coastal Flooding Hazards Based on Storm and Sea Level Rise Scenarios: Implications for Adaptation Networks in the San Francisco Bay Region." *Regional Environmental Change* 18(5): 1343–1355.

IPCC. 2014. *AR5 Climate Change 2014: Impacts, Adaptation, and Vulnerability – IPCC*. IPCC.

Isenberg, Phillip L. 2016. "Commentary: Public Policy is Messy: Three Studies in Water Management." *Public Administration Review* 76(5): 751–752.

Javeline, Debra. 2014. "The Most Important Topic Political Scientists are Not Studying: Adapting to Climate Change." *Perspectives on Politics* 12(2): 420–434.

Jenkins-Smith, Hank C., and Paul A. Sabatier. 1994. "Evaluating the Advocacy Coalition Framework." *Journal of Public Policy* 14(2): 175–203.

Kammerer, Marlene et al. 2021. "What Explains Collaboration in High and Low Conflict Contexts? Comparing Climate Change Policy Networks in Four Countries." *Policy Studies Journal* 49(4): 1065–1086.

Keohane, Robert O., and David G. Victor. 2011. "The Regime Complex for Climate Change." *Perspectives on Politics* 9(1): 7–23.

Khan, Mizan et al. 2020. "Twenty-Five Years of Adaptation Finance through a Climate Justice Lens." *Climatic Change* 161(2): 251–269.

Koski, Chris, and Alma Siulagi. 2016. "Environmental Harm or Natural Hazard? Problem Identification and Adaptation in U.S. Municipal Climate Action Plans." *Review of Policy Research* 33(3): 270–290.

Krivitsky, Pavel N. et al. 2011. "Adjusting for Network Size and Composition Effects in Exponential-Family Random Graph Models." *Statistical Methodology* 8(4): 319–339.

Leach, William D., and Paul A. Sabatier. 2005. "To Trust an Adversary: Integrating Rational and Psychological Models of Collaborative Policymaking." *American Political Science Review* 99(4): 491–503.

Lee, Kai N. 2001. "Appraising Adaptive Management." In *Biological Diversity*, CRC Press.

Leifeld, Philip. 2013. "Reconceptualizing Major Policy Change in the Advocacy Coalition Framework: A Discourse Network Analysis of German Pension Politics." *Policy Studies Journal* 41(1): 169–198.

Leifeld, Philip. 2020. "Policy Debates and Discourse Network Analysis: A Research Agenda." *2020* 8(2): 4.

Levy, Michael A., and Mark N. Lubell. 2017. "Innovation, Cooperation, and the Structure of Three Regional Sustainable Agriculture Networks in California." *Regional Environmental Change*.

Little, R. J. A., and D. B. Rubin. 2019. *Statistical Analysis with Missing Data*. Wiley.

Long, Norton E. 1958. "The Local Community as an Ecology of Games." *American Journal of Sociology* 64(3): 251–261.

Lubell, M. et al. 2019. *The Governance of Sea Level Rise in the San Francisco Bay Area: Results from a Survey of Stakeholders*. University of California, Davis.

Lubell, Mark. 2004a. "Collaborative Watershed Management: A View from the Grassroots." *Policy Studies Journal* 32(3): 341–361.

Lubell, Mark. 2004b. "Resolving Conflict and Building Cooperation in the National Estuary Program." *Environmental Management* 33(5): 677–691.

Lubell, Mark. 2014. "Network Structure and Institutional Complexity in an Ecology of Water Management Games." *Ecology and Society* 19(4).

Lubell, Mark. 2016. "Transaction Costs and the Perceived Effectiveness of Complex Institutional Systems." *Public Administration Review*: n/a-n/a.

Lubell, Mark. 2017. *The Governance Gap: Climate Adaptation and Sea-Level Rise in the San Francisco Bay Area*. University of California, Davis.

Lubell, Mark. 2020. "The Origins of Conflict in Polycentric Governance Systems." *Public Administration Review* 80(2): 222–233.

Lubell, Mark. 2021. "Collective Action Problems and Governance Barriers to Sea-Level Rise Adaptation in San Francisco Bay." *Climatic Change* 167(3): 46.

Lubell, Mark. 2022. "Methodological Approaches to the Ecology of Games Framework." In *Methods of the Policy Process*, Routledge.

Maag, Simon, and Manuel Fischer. 2018. "Why Government, Interest Groups, and Research Coordinate: The Different Purposes of Forums." *Society & Natural Resources*: 1–18.

Madanat, Samer Michel et al. 2019. "The Benefits of Cooperative Policies for Transportation Network Protection from Sea Level Rise: A Case Study of the San Francisco Bay Area." *Transport Policy* 76: A1–9.

Manfreda, Katja Lozar, Michael Bosnjak, Jernej Berzelak, Iris Haas, et al. 2008. "Web Surveys versus Other Survey Modes: A Meta-Analysis Comparing Response Rates." *International Journal of Market Research* 50(1): 79–104.

McDonald, Rachel I. et al. 2015. "Personal Experience and the 'Psychological Distance' of Climate Change: An Integrative Review." *Journal of Environmental Psychology* 44: 109–118.

McGinnis, Michael D., and Elinor Ostrom. 2012. "Reflections on Vincent Ostrom, Public Administration, and Polycentricity." *Public Administration Review* 72(1): 15–25.

McPherson, Miller et al. 2001. "Birds of a Feather: Homophily in Social Networks." *Annual Review of Sociology* 27(1): 415–444.

Meadows, Robin. 2021. "Regional Planning for Sea-Level Rise is Key to Environmental Justice." *Bay Area Monitor*.

Meier, Kenneth J., and Laurence J. O'Toole Jr. 2013. "Subjective Organizational Performance and Measurement Error: Common Source Bias and Spurious Relationships." *Journal of Public Administration Research and Theory* 23(2): 429–456.

Meijerink, Sander, and Sabina Stiller. 2013. "What Kind of Leadership Do We Need for Climate Adaptation? A Framework for Analyzing Leadership Objectives, Functions, and Tasks in Climate Change Adaptation." *Environment and Planning C: Government and Policy* 31(2): 240–256.

Mewhirter, Jack, Mark Lubell, and Ramiro Berardo. 2018. "Institutional Externalities and Actor Performance in Polycentric Governance Systems." *Environmental Policy and Governance* 28(4): 295–307. https://doi.org/10.1002/eet.1816.

Mewhirter, Jack, and Ramiro Berardo. 2019a. "The Impact of Forum Interdependence and Network Structure on Actor Performance in Complex Governance Systems." *Policy Studies Journal* 47(1): 159–177.

Mewhirter, Jack, and Ramiro Berardo. 2019b. "The Impact of Forum Interdependence and Network Structure on Actor Performance in Complex Governance Systems." *Policy Studies Journal* 47(1): 159–177.

Morrison, T. H. et al. 2019. "The Black Box of Power in Polycentric Environmental Governance." *Global Environmental Change* 57: 101934.

Morrison, Tiffany H. et al. 2023. "Building Blocks of Polycentric Governance." *Policy Studies Journal* n/a(n/a).

Moser, Susanne C., and Julia A. Ekstrom. 2010. "A Framework to Diagnose Barriers to Climate Change Adaptation." *Proceedings of the National Academy of Sciences* 107(51): 22026–22031.

Ostrom, Elinor. 1990. Governing the Commons: The Evolution of Institutions for Collective Action. Political Economy of Institutions and Decisions. Cambridge: Cambridge University Press.

Ostrom, Elinor. 2010. "A Long Polycentric Journey." *Annual Review of Political Science* 13(1): 1–23.

Ostrom, Elinor. 2010. "Beyond Markets and States: Polycentric Governance of Complex Economic Systems." *American Economic Review* 100(3): 641–672.

Ostrom, Elinor. 2010. "Polycentric Systems for Coping with Collective Action and Global Environmental Change." *Global Environmental Change* 20(4): 550–557.

Ostrom, Vincent et al. 1961. "The Organization of Government in Metropolitan Areas: A Theoretical Inquiry." *American Political Science Review* 55(4): 831–842.

Paavola, Jouni. 2008. "Science and Social Justice in the Governance of Adaptation to Climate Change." *Environmental Politics* 17(4): 644–659.

Pahl-Wostl, Claudia. 2009. "A Conceptual Framework for Analysing Adaptive Capacity and Multi-Level Learning Processes in Resource Governance Regimes." *Global Environmental Change* 19(3): 354–365.

Pasquier, Ulysse et al. 2020. "'We Can't Do It on Our Own!': Integrating Stakeholder and Scientific Knowledge of Future Flood Risk to Inform Climate Change Adaptation Planning in a Coastal Region." *Environmental Science & Policy* 103: 50–57.

Putnam, Robert D. et al. 1993. *Making Democracy Work: Civic Traditions in Modern Italy.* Princeton, NJ: University Press.

Ready, Elspeth, and Eleanor Power. 2018. "'ERGM Predictions and GWESP.' Social Network Analysis for Anthropologists."

Schafer, Joseph L., and John W. Graham. 2002. "Missing Data: Our View of the State of the Art." *Psychological Methods* 7(2): 147–177. https://doi.org/10.1037/1082-989X.7.2.147.

Schlager, Edella. 1995. "Policy Making and Collective Action: Defining Coalitions within the Advocacy Coalition Framework." *Policy Sciences* 28(3): 243–270.

Shilling, Fraser M. et al. 2016. "Adaptive Planning for Transportation Corridors Threatened by Sea Level Rise." *Transportation Research Record* 2599(1): 9–16.

Small, Mario L. 2009. "'How Many Cases Do I Need?': On Science and the Logic of Case Selection in Field-Based Research." *Ethnography* 10(1): 5–38.

Small, Mario L., and Jenna M. Cook. 2021. "Using Interviews to Understand Why: Challenges and Strategies in the Study of Motivated Action." *Sociological Methods & Research*: 004912412199555.

Smith, Jeffrey A., and James Moody. 2013. "Structural Effects of Network Sampling Coverage I: Nodes Missing at Random." Social Networks 35(4): 652–68.

Stacey, M. T. et al. 2017. "Regional Interdependence in Adaptation to Sea Level Rise and Coastal Flooding."

Suarez-Alvarez, Maria M. et al. 2012. "Statistical Approach to Normalization of Feature Vectors and Clustering of Mixed Datasets." *Proceedings of the Royal Society A: Mathematical, Physical and Engineering Sciences* 468-(2145): 2630–2651.

Teodoro, Jose Daniel et al. 2021. "Quantifying Stakeholder Learning in Climate Change Adaptation across Multiple Relational and Participatory Networks." *Journal of Environmental Management* 278: 111508.

Tompkins, Emma L., and Hallie Eakin. 2012. "Managing Private and Public Adaptation to Climate Change." *Global Environmental Change* 22(1): 3–11.

Ulibarri, Nicola et al. 2023. "Drivers and Dynamics of Collaborative Governance in Environmental Management." *Environmental Management* 71(3): 495–504.

Vantaggiato, Francesca Pia et al. 2023. "Creating Adaptive Social–Ecological Fit: The Role of Regional Actors in the Governance of Sea-Level Rise Adaptation in San Francisco Bay." *Global Environmental Change* 80: 102654.

Vantaggiato, Francesca Pia, and Mark Lubell. 2020. *Learning to Collaborate: Lessons Learned from Governance Processes Addressing the Impacts of Sea Level Rise on Transportation Corridors Across California.*

Vantaggiato, Francesca Pia, and Mark Lubell. 2022. "The Benefits of Specialized Knowledge in Polycentric Governance." *Policy Studies Journal* 50(4): 849–876.

Vantaggiato, Francesca Pia, and Mark Lubell. 2023. "Functional Differentiation in Governance Networks for Sea Level Rise Adaptation in the San Francisco Bay Area." *Social Networks* 75: 16–28.

Vignola, Raffaele et al. 2017. "Leadership for Moving the Climate Change Adaptation Agenda from Planning to Action." *Current Opinion in Environmental Sustainability* 26–27: 84–89.

Vogel, D. 2018. *"California Greenin": How the Golden State Became an Environmental Leader.* Princeton NJ: Princeton University Press.

Weible, Christopher M. et al. 2020. "Portraying the Structure and Evolution of Polycentricity via Policymaking Venues." *International Journal of the Commons* 14(1): 680–691.

Weible, Christopher M., and Tanya Heikkila. 2017. "Policy Conflict Framework." *Policy Sciences* 50(1): 23–40.

Woodruff, Sierra C. et al. 2020. "Is Coastal Adaptation a Public Good? The Financing Implications of Good Characteristics in Coastal Adaptation." *Journal of Environmental Planning and Management*: 1–20.

Yang, Jaewon, and Jure Leskovec. 2012. "Community-Affiliation Graph Model for Overlapping Network Community Detection." In 2012 IEEE 12th International Conference on Data Mining, 1170–1175. https://doi.org/10.1109/ICDM.2012.139.

Yi, Hongtao. 2018. "Network Structure and Governance Performance: What Makes a Difference?" *Public Administration Review* 78(2): 195–205.

Young, Oran. 2006. "Vertical Interplay among Scale-Dependent Environmental and Resource Regimes." *Ecology and Society* 11(1).

Cambridge Elements ≡

Organizational Response to Climate Change

Aseem Prakash

University of Washington

Aseem Prakash is Professor of Political Science, the Walker Family Professor for the College of Arts and Sciences, and the Founding Director of the Center for Environmental Politics at University of Washington, Seattle. His recent awards include the American Political Science Association's 2020 Elinor Ostrom Career Achievement Award in recognition of "lifetime contribution to the study of science, technology, and environmental politics," the International Studies Association's 2019 Distinguished International Political Economy Scholar Award that recognizes "outstanding senior scholars whose influence and path-breaking intellectual work will continue to impact the field for years to come," and the European Consortium for Political Research Standing Group on Regulatory Governance's 2018 Regulatory Studies Development Award that recognizes a senior scholar who has made notable "contributions to the field of regulatory governance."

Jennifer Hadden

University of Maryland

Jennifer Hadden is Associate Professor in the Department of Government and Politics at the University of Maryland. She conducts research in international relations, environmental politics, network analysis, nonstate actors, and social movements. Her research has been published in various journals, including the *British Journal of Political Science, International Studies Quarterly, Global Environmental Politics, Environmental Politics,* and *Mobilization.* Dr. Hadden's award-winning book, *Networks in Contention: The Divisive Politics of Global Climate Change,* was published by Cambridge University Press in 2015. Her research has been supported by a Fulbright Fellowship, as well as grants from the National Science Foundation, the National Socio-Environmental Synthesis Center, and others. She held an International Affairs Fellowship from the Council on Foreign Relations for the 2015–16 academic year, supporting work on the Paris Climate Conference in the Office of the Special Envoy for Climate Change at the US Department of State.

David Konisky

Indiana University

David Konisky is Professor at the Paul H. O'Neill School of Public and Environmental Affairs, Indiana University, Bloomington. His research focuses on US environmental and energy policy, with particular emphasis on regulation, federalism and state politics, public opinion, and environmental justice. His research has been published in various journals, including the *American Journal of Political Science, Climatic Change,* the *Journal of Politics, Nature Energy,* and *Public Opinion Quarterly.* He has authored or edited six books on environmental politics and policy, including *Fifty Years at the U.S. Environmental Protection Agency: Progress, Retrenchment and Opportunities* (Rowman & Littlefield, 2020, with Jim Barnes and John D. Graham), *Failed Promises: Evaluating the Federal Government's Response to Environmental Justice* (MIT, 2015), and *Cheap and Clean: How Americans Think about Energy in the Age of Global Warming* (MIT, 2014, with Steve Ansolabehere). Konisky's research has been funded by the National Science Foundation, the Russell Sage Foundation, and the Alfred P. Sloan Foundation. Konisky is currently coeditor of *Environmental Politics.*

Matthew Potoski

UC Santa Barbara

Matthew Potoski is a Professor at UCSB's Bren School of Environmental Science and Management. He currently teaches courses on corporate environmental management, and his research focuses on management, voluntary environmental programs, and public policy. His research has appeared in business journals such as *Strategic Management Journal, Business Strategy and the Environment,* and the *Journal of Cleaner Production,* as well as public policy and management journals such as *Public Administration Review* and the *Journal of Policy Analysis and Management.* He coauthored *The Voluntary Environmentalists* (Cambridge, 2006) and *Complex Contracting* (Cambridge, 2014; the winner of the 2014 Best Book Award, American Society for Public Administration, Section on Public Administration Research) and was coeditor of *Voluntary Programs* (MIT, 2009). Professor Potoski is currently coeditor of the *Journal of Policy Analysis and Management* and the *International Public Management Journal.*

About the Series

How are governments, businesses, and nonprofits responding to the climate challenge in terms of what they do, how they function, and how they govern themselves? This series seeks to understand why and how they make these choices and with what consequence for the organization and the eco-system within which it functions.

Cambridge Elements ☰

Organizational Response to Climate Change

Elements in the Series

Explaining Transformative Change in ASEAN and EU Climate Policy: Multilevel Problems, Policies and Politics
Charanpal Bal, David Coen, Julia Kreienkamp, Paramitaningrum and Tom Pegram

Fighting Climate Change through Shaming
Sharon Yadin

Governing Sea Level Rise in a Polycentric System: Easier Said than Done
Francesca Pia Vantaggiato and Mark Lubell

A full series listing is available at: www.cambridge.org/ORCC

Printed in the United States
by Baker & Taylor Publisher Services